Primary Science

Primary Science

TEACHING THE TRICKY BITS

G. N. Rutledge

Open University Press
McGraw-Hill Education
McGraw-Hill House
Shoppenhangers Road
Maidenhead
Berkshire
England
SL6 2QL

email: enquiries@openup.co.uk
world wide web: www.openup.co.uk

and Two Penn Plaza, New York, NY 10121–2289, USA

First published 2010

Reprinted 2012

A catalogue record of this book is available from the British Library

ISBN-13: 978 0 335 22228 5 (pb) 978 0 335 22229 2 (hb)
ISBN-10: 0 335 22228 5 (pb) 0 335 22229 3 (hb)

Library of Congress Cataloging-in-Publication Data
CIP data applied for

Typeset by RefineCatch Limited, Bungay, Suffolk
Printed in the UK by Bell and Bain Ltd, Glasgow

The McGraw-Hill Companies

Contents

Acknowledgements

Learning is a collaborative, dialectical process. The learning in this book has been thanks to children, teachers and colleagues too numerous to list but their vital assistance is freely acknowledged.

Especial thanks to Joanne Hills and Simon Wilkinson for help with illustrations and many other tasks.

1

Introduction

For me, one of the great attractions of being a primary school teacher is the fact that there is such a spread of interesting subjects to get to grips with. Of course this can also be unnerving, as no one is going to be an expert in all the subjects and all of us will have some subjects where we feel less confident. Many teachers feel this way about primary science and often become quite intimidated by the subject knowledge they are required to help children learn. This might seem strange as there are lots of accessible publications designed to provide the necessary background knowledge as well as numerous schemes designed to provide outlines of suitable learning activities. Why, even with these resources, do so many primary school teachers lack confidence in the teaching of primary science? Having spent years working with students and teachers across the UK, it seems to me that the problem lies in the gap between these two sorts of materials: the subject knowledge books provide lists of factual material; the schemes provide ideas for activities; but in between there is a gap that might be defined by the question, 'What sort of understanding am I aiming for and how do I use this activity to get it across to the children?' Other related questions I have often had teachers ask me include: 'What should I tell the children and what should I try to let them discover on their own?'; 'What vocabulary and models should I use to explain these concepts?'; 'What's the right sort of level of understanding for a year two child?'; or 'How should the understanding of the year sixes in my composite class differ from the year fours?'

The nub of the problem is that we often treat subject knowledge as lists of facts. These can be very basic facts such as, 'forces are measured in Newtons', or conceptually more complex facts such as, 'if an object is stationary, it is not that there are no forces acting on it, but that the forces are balanced'. The problem is that we do not understand our surroundings as a collection of facts, what is more important is how we link the facts together and apply them as models to interact with and understand the world we are part of. That is the nature of the gap between the two sorts of materials mentioned: it is not enough to know the facts and have ideas to help children learn them; we must also know how to help the children link these facts into models of understanding and, crucially, how to refine and progress these models as they gain in experience.

There are some good materials that address such conceptual progression and understanding of science amongst primary school children but, in my experience, teachers view them largely as 'theory books' and, in the pressure of everyday work in the classroom, find them rather inaccessible. Even student teachers often regard such texts as relevant to writing their college assignments but somehow not applicable to 'real life in the classroom'. This book attempts to fill at least part of this gap by providing accessible, practical assistance to help you encourage children to begin to link facts into progressive models of understanding that will allow them to more fully interact with and explain the universe around them.

This task is an ambitious one and there are several potential pitfalls. The next chapter on how to use the book gives advice on avoiding these.

2

How to use this book

When we are in the classroom and under pressure, especially when we are planning learning in subject areas where we are not confident, there is a great temptation to look for quick fixes. This is the perennial appeal of the likes of commercial science schemes or government exemplars. Such schemes can appear a quick way of providing us with programmes of learning that we can use as the basis of our planning. Pressurized as we are, many of us fall into the trap of lifting such schemes and using them unmodified to provide for the learning of our classes. The schemes are often very useful as a stimulus for our planning but every school, every class and every child is different and these facts alone should warn us that employing such schemes uncritically and without customization will erode learning. This book is no different. The ideas it contains are the fruit of many years managing science in primary schools, delivering continuing professional development and working with students in initial teacher education, but they are not a magic answer to every problem. All of the approaches discussed have been used successfully in a wide range of primary schools across Northern Ireland, Scotland and England but this doesn't mean that they can be slavishly applied in every context. Accordingly it is important to bear the following points in mind.

Take care when using the book to remember:

1 The book aims to help you ensure effective conceptual progress for your learners. This means that it is largely focused on what is usually termed science knowledge and understanding, as opposed to science skills. This is not to suggest that science skills are unimportant, indeed they are vital, and even if skills are not the main emphasis of this book, they are embedded in its core principle. This core principle is the idea that we as teachers should not just be telling our learners facts but instead should find out their ideas and then allow them the chance to challenge and test these ideas. This manner of working is impossible without science skills and Chapter 5 discusses this further.
2 Even within the field of science knowledge and understanding, the book does not cover every aspect of primary science; instead it concentrates on those areas where

experience and research show practitioners and learners to have the greatest difficulties. There are some topics where it is relatively easy for even an inexperienced teacher to prepare sound learning experiences and ensure good conceptual progress. There are other areas where even a diligent and experienced practitioner can fall foul of the confusion and misconceptions that result from the wonderful, yet counter-intuitive universe that we live in. It makes sense therefore to concentrate on the tricky bits; hence the title of this book!

3 For each area covered, advice is given to help you facilitate children in developing models that will aid them in their understanding of the world around them. Each chapter is designed to show typical conceptual progression for the topic concerned. The progression covered illustrates that from around the age of 5 to the age of 10 or 11, the age range covered by key stages 1 and 2 in the various UK curricula. Of course every child is an individual and their understanding may well not be typical of their age, or even normal progression, but the outlines will give you a solid foundation on which to base planning for learning in the most difficult topics. At all times you should view the ideas critically and modify them to suit the circumstances of your particular learners.

4 The materials concentrate on giving advice on progressive models of understanding but they also give ideas for activities to help the children develop these. The suggested activities will go a long way to helping you with your planning for these topics but they are not exhaustive and you will need to supplement them. This will be essential for those more straightforward aspects of science topics that this book does not address.

5 The core of the approaches given is that they are centred on the children's ideas. This creates a paradox: how can I give advice to assist you in helping your children learn, when I do not know the specific ideas held by your children? The good news is that, whilst every child is different, there are common patterns of understanding encountered. This means that experience, and careful consideration of the research into children's understanding, allows us to plan ahead with confidence regarding the likely difficulties we will face. A warning is necessary however! The materials given here reflect the common understanding and ideas of children but you will encounter children who do not fit the typical patterns. Accordingly it is important not to apply the ideas in the book slavishly but to be prepared to respond flexibly to your learners' needs. You needn't panic about this however as, if you adopt the approaches outlined, you will find out exactly what your children think and will therefore be able respond appropriately.

So, the book is not a complete answer to all needs regarding the management of learning in primary science but then such a resource is impossible! There are no quick fixes for learning and theorists can sometimes be scathing about what they disparagingly term, 'tips for teachers'. Don't be perturbed by this though, an important part of learning is to benefit from others' experience and, if you employ this book critically alongside the materials and schemes your school already uses, you will be able to enhance your children's learning, particularly by ensuring progression in their conceptual understanding.

Features of the book

After this chapter there are a further three background chapters:

- Chapter 3: How children learn;
- Chapter 4: Ensuring an effective learning context;
- Chapter 5: The relationship between science skills and knowledge and understanding.

These chapters set out important aspects of practice as well as the philosophy behind the book, so it is important that you get to grips with the principles they put forward before using the later chapters.

After these opening chapters, the book is divided into topic areas that reflect the most difficult aspects of primary science learning. Each of these chapters has a common format consisting of the following elements:

- *A general introduction* This notes important elements of the likely context of the learning.
- *Where the topic fits in* This gives an outline of where the content is likely to fit into the broader curriculum.
- *Links to other topics* This section will show you where links need to made to other topics in the book.
- *A context for the topic* For each topic, an example learning context is given based on the principles outlined in Chapter 4, Ensuring an effective learning context.
- *A step-by-step contextual progression* The learning for the topic is presented as a series of conceptual steps arranged in the order best suited to supporting the children's learning. As part of this section, there is a table listing the conceptual steps that are likely to prove most useful for the topic. Beside the conceptual steps, the tables give examples of the learning objectives that would be necessary if the children were to successfully achieve that conceptual step. Work with students and practising teachers suggests that many of you will find these lists particularly useful. If you are lacking in confidence and experience it can be hard to know how to break down the children's learning into focused, progressive steps and unfortunately the various national curricula are rarely much help in this respect. This is because the curriculum statements, although often not inappropriate, are broad statements of learning which need to be further broken down into learning objectives. Another common problem is that the national curricula statements are not always listed in the most useful order to ensure effective conceptual progression. The suggested learning objectives are not tied to any particular set of curriculum statements and are formulated to support a more effective progression. This means that, whatever curriculum you are working under, it should be possible to use the guidance to improve your children's learning.
- As usual though, there is a warning! The example learning objectives are based on years of successful practice with a wide range of children and represent very

useful scaffolding. But, once again, remember: every child, and every teacher, is different and so you should base your work on your learners' own ideas and be prepared to be flexible.

- *Eliciting ideas* Often each conceptual step will require you to find out what your children think about the concepts concerned and advice on how to do this is given.
- *Be warned!* These sections alert you to common misconceptions held by children and other difficulties you are likely to encounter.
- *Challenging the children's ideas* You must be prepared to challenge the children's likely misconceptions and so advice on how to do this is provided.
- *Hint!* These sections give tips on strategies that have proved particularly useful in supporting learning.

Getting started

Before we attend to particular topics, it is important to consider how children actually learn. This will allow us to establish an approach that will maximize their progress. Such issues are examined in the next two chapters. All of the strategies outlined in this book are based on the philosophies explored in these chapters and so it is important to consider them carefully.

3

How children learn

All the approaches in this book are based on what is termed a constructivist approach to learning and it is important to understand how its philosophy underpins and influences the strategies outlined.

An outline of constructivism as relevant to primary science

Constructivism holds that the major influence on children's learning is their prior experience, which results in their developing *constructs* that govern how they interpret and relate to the world around them. These constructs are almost always subconscious but they are always present. For example, even if a child says, 'I have no idea why things fall', they will have some ideas; they just don't consciously articulate them.

These constructs are formed from a very early age. Just because a key stage 1 child has never had any lessons on forces does not mean that they will not have ideas about why things fall. In this example such young children typically hold the scientifically incorrect belief that 'things just fall'; they do not appreciate that it takes a force to make objects fall; in this case, gravity.

This idea, 'that things just fall', may be incorrect but it is certainly not a foolish idea; it is easy to appreciate why young children hold it and why it makes sense to them. Such scientifically incorrect yet sensible ideas are termed *misconceptions*. You will find that it is very difficult to change children's misconceptions. The main reason for this is that the misconceptions often make more sense to the children than the scientifically correct ideas do! This means that often the children either subconsciously, or consciously, reject the scientific explanation.

When you are attempting to change children's misconceptions, the key point to realize is that simply telling the children the correct answer is most unlikely to work. If children are to modify their ideas, they must themselves accept that their initial ideas were incorrect and then build an alternative, scientifically correct, construct to explain the phenomena they are encountering. To do this, the children's misconceptions must first be identified and the reasons why they hold them understood. The misconceptions can then be challenged, preferably by the children themselves testing them, and finally the appropriate ideas can be consolidated.

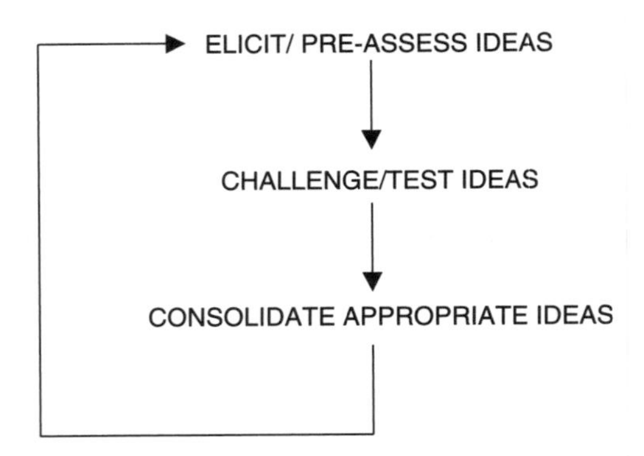

Figure 3.1 A learning cycle for primary science

This means that it is important that you plan for the children's learning cyclically, as shown in Figure 3.1.

An example of the learning cycle in action

It is perhaps easier to understand how the cycle works by extending the above example.

Some years ago I was working in school with a class of nine-year-olds who were to learn about how objects fall as part of their work on forces. The first stage was to *elicit and pre-assess the children's ideas* on the subject. This was done by giving groups of children a wide variety of objects to drop and asking them to make predictions about how the objects would fall. The children then dropped the objects and compared their observations with their predictions. I then talked to each group of children, asking them to explain their ideas and why they thought objects fell. All the children stated this was because of gravity. The children had obviously been told this in earlier science lessons but it was important to further probe their understanding as to exactly what gravity was. When questioned on this they typically made statements such as, 'Gravity's the force that pulls things to Earth'. Already, therefore, I was receiving warnings about the children's understanding as it appeared that some thought gravity was a force only found on the Earth; a common misconception. Experience had taught me that it was necessary to probe yet further as to what the children thought caused gravity. I did this and was not surprised to find that, without exception, the children believed that gravity was caused by air pushing the objects down, once again a very common misconception. In no way was this idea a stupid one, as an example of the children's reasoning will illustrate. A significant number of children (close to a quarter of the class) told me that there was no gravity in space because there was no air. Several of these children then backed up their judgement by explaining that that was why astronauts had to wear space suits to be able to breathe. This was sophisticated reasoning, even if scientifically incorrect, and deserved careful attention. (I will further discuss just how sophisticated one child's ideas were in Chapter 19.)

Now, having elicited the children's ideas (which in this case proved to be misconceptions) the next step was to *challenge/test the children's ideas*. The clear way to do this was to find examples of where objects fell without air. Sometimes the children can carry out an investigation to allow them to test their ideas but in this case this was not possible, there being no means available for children to drop objects in a vacuum. The ideas still had to be challenged however; it was not enough for me to 'tell the children the right answer'. They had obviously been 'told' about gravity before but had not had their ideas thoroughly explored and challenged. If the children were to learn effectively, they would have to connect up with their learning and 'change their own minds'. This was not difficult to arrange and was achieved by a combination of thought experiments and video evidence. First we discussed the Moon and the children quickly decided there was no air there as astronauts needed suits to breathe. The children were then asked if they thought there was gravity on the Moon and they decided no, as there was no air. They were then asked to predict if objects would fall on the Moon and the children typically made statements such as, 'No, they would just float about'. It was then a simple matter to access, through the Internet, clips of astronauts jumping on the Moon and falling back to the surface and dropping objects, which clearly fell to the ground. The challenge was then taken further with the children being asked that, if gravity was caused by air, would humans be able to breathe where there was no gravity? After watching the first clips the children were already changing their minds about their original theory but they agreed that, if their first theory was correct, humans would not be able to breathe where there was no gravity, as there would be no air. We then watched clips of astronauts floating in zero gravity in space stations but not wearing space suits; clearly there was air but no gravity. The children were all now convinced that gravity could not be caused by air.

After this it was a relatively simple matter to design activities that *consolidated the appropriate ideas* and more advice on this is given in Chapter 19.

The importance of the learning cycle

Working with your children like this is not difficult once you get into the habit of doing so. Still, it can take some getting used to as we have often got into the habit of 'teaching' the children rather than 'helping them learn'. Often many of the activities we present the children with end up being used only to reinforce what we are already 'telling' the children. Such an approach does not allow the children to access their ideas, seriously consider their opinions and then modify them where necessary. This is a key point about the *eliciting/pre-assessing children's ideas* chapter of the cycle. Eliciting children's ideas is important not just to warn us of misconceptions that the children may hold but it is equally crucial as it allows the children to become aware of what they are thinking and why. In the example above, the learning was effective as the children *themselves* realized what they believed had to be incorrect. Had I simply told them that they were wrong and provided an alternative explanation, the likelihood is that, for many children, the learning would not have been embedded and they would have subconsciously reverted to their previous, and to them, more logical beliefs. Misconceptions are notoriously hard to change and children will certainly not modify their ideas without them being properly challenged in a targeted manner, hence the importance of the *challenge/test ideas* stage. Once these two stages have been properly addressed the *consolidating appropriate ideas* stage is relatively straightforward. The problem is that we often start off at this stage without having gone through the first two steps. The results can be seen by working with any group of postgraduate or undergraduate initial teacher education students: they will all have covered the science topics addressed, at least up to the age of 14 or so, yet they routinely hold exactly the same misconceptions held by primary children. Why? Because they have often just 'learnt from the book' and, not having had their ideas adequately challenged, they revert to the misconceptions that their 'common sense' favours over the correct, but often counter-intuitive, scientific explanations.

This book will help you with using the learning cycle approach with your children as the various chapters give suggestions for how to elicit the children's ideas and also warn you of particularly common misconceptions. But by now the importance of the warnings in the last chapter, regarding how the book cannot be seen as a complete answer, and how it should not be followed slavishly, should be clear. You never know exactly what your children are going to think but, as I said earlier, the good news is that they tend to think in a predictable fashion and so the materials that follow should prove of considerable assistance, so long as you are careful to use them flexibly to suit the particular needs of your learners.

Open-ended and discovery learning

It is often suggested that children will learn more effectively if they are allowed to approach their learning in an open-ended and exploratory fashion. This is true to a point, but there are also serious potential pitfalls if the approach to learning is too unstructured. Open-ended, exploratory approaches are often a great way to address the *elicit/pre-assess ideas* part of the learning cycle, as they encourage the children to reflect on what they believe. They are also a good way of generating ideas that can then be *challenged/tested*. An example of such an approach is the opening activity described above, where the children were given a wide variety of different objects to explore how they fell. The children made a variety of predictions and observations and came up with a series of factors that they felt might influence how objects fell such as shape, surface area and weight. It was then easy for the children to test their ideas in a series of investigations.

The children were able to have fun carrying out the tests and also had a greater sense of achievement as they had ownership of the learning; they were devising *their own* tests to test *their own* ideas (albeit with guidance). At the same time, however, we must recognize that others have worked to create a scientifically correct model of gravity and falling. We are most unlikely in the primary classroom to produce evidence to challenge the scientific view of gravity and falling that has been built up and modified over the years by figures such as Galileo, Newton and Einstein. We must be careful therefore to ensure we are helping the children work towards an understanding of these accepted scientific models. Sometimes I have encountered an attitude that the most important thing is to let the children develop their own ideas, at 'their own level' and that, even if these ideas are scientifically incorrect, they will somehow get 'sorted out in time'. But, as we have seen from the discussion above, children will always develop their own ideas, which will often include misconceptions that are extremely resistant to 'sorting out'.

It seems to me therefore that we should certainly centre our work on our learners' ideas but that we should be constantly striving to allow them to challenge and test their ideas *within a framework* that will ensure they progress within scientifically correct boundaries. An open-ended approach to learning can have disastrous results in this respect. For instance if, when learning about electricity, we allow the children to build circuits containing differing appliances, for example a bulb and a buzzer, they will be faced with baffling results. In this case, assuming a 4.5V battery, a 1.5V bulb and a 3V buzzer, the buzzer will buzz but the bulb will not light. How will the children explain this? They will certainly have no difficulty coming up with ideas but they are most likely to be along the lines of, 'the buzzer uses most of the electricity and so there's not enough left to light the bulb'. This is a seemingly sensible suggestion but it is, of course, scientifically completely incorrect. And what of the correct explanation? It is closer to A-level standard than primary school science. I make it a guiding principle that, as far as possible, the children should not be introduced to concepts that they do not have a fighting chance of being able to explain at a level accessible to them. In this case, it is better to have a more structured approach rather than the children ending up confused. In fact a reasonable primary school explanation of this phenomenon can be attained but it takes experience, carefully planned learning and confidence on the part of the teacher. The issues are discussed in Chapter 17.

All this can seem a bit daunting but the book will help you recognize when an open-ended approach is suitable and when you would be well advised to use more structured activities.

4

Ensuring an effective learning context

As well as considering the points about how children learn given in the preceding chapter, it is important to realize that the context of your children's learning has a major influence on how effectively they learn and that it is therefore vital to understand what makes for a good learning context.

There are five areas that I consider when planning learning contexts.

- Is the context fun? Will it motivate and enthuse the learners?
- Is the context accessible? Will it make sense to the learners?
- Is the context relevant? Does it provide the learners with a reason to carry out their work?
- Is the context appropriate? Does it support the necessary progression of skills and concepts?
- Is the context flexible? Will it support differentiation and a cross-curricular approach?

A fun learning context

It should really go without saying that you should plan learning for your children that they will enjoy. If your children are enjoying themselves and are well motivated, they will learn better; it is as simple as that. But creativity and imagination put into planning your learning contexts will bring other dividends too. The chief benefit that I have found is that happy, motivated children are much easier to manage. Less time spent on behavioural issues will mean more time spent on learning, as well as reducing everybody's stress levels.

An accessible learning context

By accessible, I mean a context that the children can relate to. For example young children are very egocentric, thus, when beginning work on living things, it is best to start with the children themselves, as that is what they will know best. Bacteria could also illustrate the principles of living organisms but these would not be accessible to the young children; they would not have had direct, conscious experience of bacteria and would find it difficult even to see them. For such children, bacteria would make an inaccessible learning context.

A relevant learning context

One of the easiest things you can do for your learners is to ensure they have a reason for whatever task they are carrying out. When monitoring student teachers' practice on placement, I always ask the children they are teaching why they think they are doing the activity they are carrying out. Often the answers are amusing but not necessarily encouraging; responses such as, 'I don't know, ask her', or, 'Because this is school and you have to do this sort of thing', crop up too frequently. Give your children a reason for their work!

Again the importance of basing your planning on the children's ideas comes through; I much prefer to hear answers such as, 'We were puzzled that the water in puddles seems to disappear and we're trying to find out where it goes'. You also have a great opportunity in the fact that primary school children have a wonderful capacity to suspend disbelief. This means that you can have all sorts of imaginative reasons for learning activities. These reasons may be make-believe but they will be accepted by the children and will motivate and enthuse them. An answer such as, 'We're trying to find the best material to keep the space drive cool so it doesn't explode', is also music to my ears.

An appropriate learning context

This one is a bit more boring but it is important to bear in mind. You might have a fun, accessible and relevant learning context but it must also support the appropriate conceptual progression for the children. For instance, when I first started teaching, I used to plan to progress the children's learning in the topic of electricity by giving them imaginative design and technology tasks. As I became more experienced, I realized that the open-ended nature of the tasks was diluting the children's learning regarding electricity and, worse still, sometimes leading to misconceptions being reinforced. I learned to plan more appropriate contexts but still found the technology topics a useful way for reinforcing and assessing the learning at a later stage.

A flexible learning context

By this I mean two things. Firstly the context must support inclusion and differentiation; all the children must be able to access their learning through the context. Secondly, I always attempt to ensure some form of cross-curricular links for my contexts. Cross-curricular aspects allow the children to make greater sense of their learning by linking up different areas of learning and seeing how they connect; the universe is not divided into subject areas! The cross-curricular aspects are also often a boon as they allow learning to be linked in what is a very crowded curriculum;

I certainly have never been able to cover all I'm supposed to with my classes without saving time by using cross-curricular contexts.

Learning contexts in this book

The purpose of this book is not to provide you with lists of possible contexts but it would be hypocritical not to address them. Accordingly, for each topic covered I've provided an example of a context that I have found meets the above criteria and has proved successful in the classroom.

5

The relationship between science skills and knowledge and understanding

The importance of science skills

It is almost universal to distinguish between *science skills* and *science knowledge and understanding* and set different learning requirements for each. The exact specification of skills varies but they include skills as diverse as predicting, choosing equipment and looking for patterns in results. It is widely acknowledged that the science skills are important and, given the discussion on how children learn, it is clear why this is so. Children will not be able to connect up with their own ideas without skills such as predicting and they will not be able to challenge their ideas without skills such as looking for patterns in results. It is vital therefore that we give proper attention to developing the children's science skills.

The role of science investigations

One application of science skills that has been much emphasized is the science investigation. The term investigation can have a variety of meanings but, in primary science, it is usually used to refer to a specific type of activity where the children test a prediction by changing one variable and measuring the effect, whilst controlling the other variables to ensure what is often called a fair test. Such investigations can indeed be a very powerful learning tool. They are a great way of combining a large spread of science skills into one activity and they are an excellent way for children to challenge/test their ideas. For example, if the children believe that wrapping objects with insulation heats them up, a common misconception, they can test their ideas by carrying out an investigation comparing insulated and non-insulated frozen objects. If their misconception was correct, the insulated frozen objects should melt faster but the investigation will quickly challenge the children's ideas and they will see that insulation actually slows the melting. (Exactly this sort of investigation is further discussed in Chapter 10.) This is a very powerful use of an investigation and an excellent illustration of their potential to challenge children's ideas. It is this capacity for challenging ideas that makes investigations a key part of the learning cycle.

The limitations of science investigations

However useful they are, investigations do have limitations. Young or inexperienced children will not be able to grasp the necessary spread of skills for a whole investigation and will have to be introduced to the various skills through other activities before they can combine them. Investigations are also often time-consuming and can be demanding to manage; meaning that experience is needed to know how and when to employ them for best effect. But perhaps the most serious limitation of investigations is that they are only possible for some aspects of primary science. Think about the example misconception regarding gravity given in Chapter 3. Would the children have been able to carry out an investigation to test if it was air that made objects fall? Such an investigation is certainly possible but not with the equipment commonly available to primary schools. Nor, indeed, was there any need for an investigation, the children's ideas could be thoroughly challenged using other activities. So, think carefully about your use of investigations, they can be a powerful learning tool but are not suitable for every situation.

The balance between science skills and knowledge and understanding

An overly narrow focus on science knowledge is rightly criticized. Such an approach leads to children being presented with limited factual knowledge which they rarely retain well and which leaves misconceptions unchallenged. Too great an emphasis on the skills also causes problems, however. Skills can only be learned effectively in context; when they are actually being used for something. A series of investigations or skills activities for their own sake will have little effect on the children's learning; they should be using the skills to explore issues and answer questions that they themselves want to find answers to. Once again, the importance of the learning cycle can be seen: if you plan in such a fashion, the children will be using and developing their skills to test and check their own ideas. This approach means much more effective learning for three main reasons: the children will connect up with their own ideas; they will have greater ownership of their learning; and the skills and knowledge aspects of the learning will be in balance.

The place of science skills in this book

Science skills are not the main subject of this book, which seeks primarily to help you develop the children's knowledge by allowing them to build up progressively more complex models to understand the world around them. But we must appreciate from the preceding discussions that the children will not be able to achieve this without developing and using their science skills. Accordingly you will see that all the topics in the book are addressed using the learning cycle summarized in Figure 3.1 and that there are constant references to how to use science skills, including investigations where appropriate, to allow the children to further their understanding.

6

Developing an initial understanding of classifying materials and their properties

Introduction

When setting out what children are to learn, science curricula usually begin by listing the science skills followed by the various aspects of subject knowledge. Often the subject knowledge sections begin with the learning associated with living organisms, followed by that to do with materials and finally the content that falls under the heading of physical processes. I am not sure why this order is chosen, although it may reflect the degree of confidence teachers express regarding the different areas; living things are undoubtedly perceived as the least problematic, whilst aspects of physical processes are always seen as the most difficult.

The order is not helpful for the children's learning, however. In the world of adult science, it is physical processes that underpin every other aspect of science and that are used to attempt to explain everything from how the molecules in a plant cell interact, to how a star generates light. In the world of primary science, however, the accessible explanations are likely to lie much more within the realm of materials and their properties and, if the children are going to be able to adequately explain what they discover in their science activities, they will need a sound grounding in this area. This means that it is a good idea to start your children off with materials topics before progressing to other areas. Of course this may not always be possible, especially in schools with composite classes and rolling curriculum programmes but it is worth striving for and the following chapters of the book will identify where knowledge of materials is essential if the children are to be able to develop adequate explanations for what they encounter.

Where this topic fits in

The content outlined in this topic should typically be covered in key stage 1. Much of it is accessible to even the least experienced children but some aspects, such as the elements to do with gases, are trickier and may prove taxing for the younger children. As always, it is sensible to pre-assess the children in later years to check that they have become secure with the content outlined. Once again we see the importance of basing our work on the children's ideas!

Links to other topics

The various national curricula usually expect the kind of concepts covered in this chapter to be extended into key stage 2, often with reference to specific properties of materials such as thermal insulation, magnetic behaviour and electrical conductivity. Such matters will be addressed in Chapters 10, 11 and 17, respectively.

A context for the topic

One of my favourite contexts for this topic involves some special dressing up and looking at the materials used in the dress and equipment of different people. The knack is to find examples of dress and equipment that incorporate a wide variety of materials; which show different properties; and that are used for various purposes. Some of the most successful examples I have used have been with the help of local police and fire officers who have been very willing to come in and discuss their specialist clothing and equipment with the children. As you can imagine, the children really enjoy this and there is scope for excellent cross-curricular extension. Your own, or friends' hobbies can also be helpful for this topic; I have successfully used my alpine climbing equipment and I love to see the children's faces when I walk into the classroom clad in my full SCUBA gear. The clothes and equipment generate lots of useful discussion, which is great for eliciting ideas. The children can then go on to activities such as labelling and classifying the materials and their properties. This can be done in discussion and, with those materials that the children can safely handle, as an independent exercise.

Contexts such as these also lead naturally to sound investigative work as the children can then test the properties of different materials with a view to equipping themselves, or fictional characters, for various expeditions. I have found a splendid way to round off the topic is to finish with such an adventure, for example a safari to a local wood where the children have to kit themselves out with the right equipment to stay safe and comfortable. They could also extend their work on materials and, appropriately supervised and with due attention to health and safety issues, complete some simple tasks such as building a shelter, and preparing their lunch.

Conceptual step one – defining what is a material

The most fundamental aspect of the topic lies in being able to understand what materials are and to be able to recognize their properties. Such themes are accessible even for the youngest children in primary schools but they are still not as straightforward as they sound! The first step is to find out the children's ideas on what they think a material is.

Table 6.1 Conceptual steps and example learning objectives for developing an initial understanding of classifying materials and their properties (all of these learning objectives are suitable for coverage at key stage 1)

Conceptual step	*Example learning objectives*
	Children will learn:
Defining what a material is.	to recognize and name commonly encountered materials; that 'material' means what something is made of;
Identifying the properties of materials.	to use their senses to notice the similarities and differences between materials; that these similarities and differences are properties of the materials;
Introducing both solids and liquids.	to be able to distinguish between solids and liquids;
Introducing common gases.	to extend the above four objectives to apply to some commonly encountered gases including being able to distinguish gases from solids and liquids;
Sorting materials.	to be able to sort objects by both material and property; to be able to correctly classify a range of common solids, liquids and gases;
Learning how the use of materials is linked to their properties.	about the uses of some common solids, liquids and gases and how these relate to their properties;
Learning where materials come from.	that materials come from a wide variety of sources and to know where some common materials come from.

Eliciting ideas

If you have adopted a similar context to that given in the example, the easiest approach is to ask the children what they think some of the equipment is made of. This will give you an insight into what materials the children can identify. You can then introduce the term *material* and explore what the children understand by it.

Be warned!

At this stage you are likely to encounter the following common misconceptions regarding what a material is:

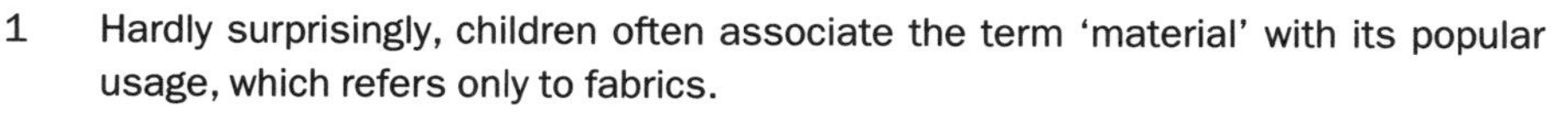

1 Hardly surprisingly, children often associate the term 'material' with its popular usage, which refers only to fabrics.

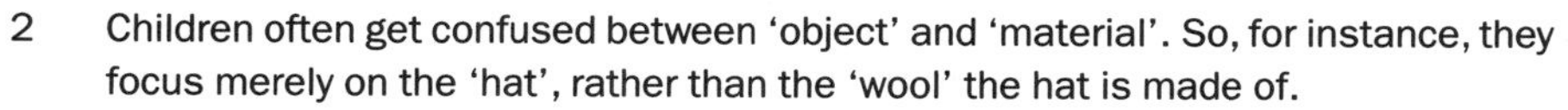

2 Children often get confused between 'object' and 'material'. So, for instance, they focus merely on the 'hat', rather than the 'wool' the hat is made of.

Challenging the children's ideas

I have found that the above misconceptions can largely be countered from the beginning by encouraging the children to always ask the question, 'What is the object made of?' and then introducing *material* as meaning *what something is made of.*

Conceptual step two – identifying the properties of materials

Following on from our dressing up context, you can now encourage the children to begin to describe and classify the materials in terms of their properties.

Be warned!

There are a number of common misconceptions that it is wise to be prepared to deal with:

1. Following on from the problem detailed above, where the children get confused between 'object' and 'material', it is not surprising that the children often describe properties of an object rather than of the material it is made of. You are likely to find your learners making statements such as, 'the ice axe is hard' rather than realizing that it is the material (metal) that it made out of that is hard.
2. Similarly, children often focus on the use of an object and describe this, rather than the properties of the material. For example children often make statements such as, 'the diving mask lets you see underwater', rather than focusing on the properties of the plastic it is made of (waterproof and transparent) that allow underwater vision.
3. A simple fact that can lead to learning being eroded in this area is that children find it easier to note differences between materials, rather than similarities.
4. A common misconception that caught me completely by surprise when I first started working with children is the fact that they find it difficult to distinguish between certain properties. The most frequent confusions I have encountered are between *smooth* and *soft* and *hard* and *strong*.

Challenging the children's ideas

The difficulties above can often be addressed by taking a few simple precautions.

Chief amongst these is to pay careful attention to the language the children use. Watch out for examples of the tell-tale misconception statements discussed. A good way of helping the children overcome these difficulties, should you encounter them, is to encourage the children to always focus on what the object is made of and to include this in their observations. Thus the potentially confusing statement, 'the ice axe is hard', becomes, 'the metal the ice axe *is made of* is hard'. This approach also helps address the second difficulty outlined, for example the unhelpful statement, 'the diving mask lets you see underwater', would then become something like, 'the plastic

The children will need initial, supported practice with such techniques at whole class or group level but they soon get used to making their choices and being ready to justify them.

Differentiation can be introduced by varying the selection of words provided. Some freedom should always be given for different choices and their interpretation and this can even be used to check for particular misconceptions. For example, here both strong and hard have been provided in the properties selection. Children often incorrectly think these are synonyms and so this exercise can check for such errors.

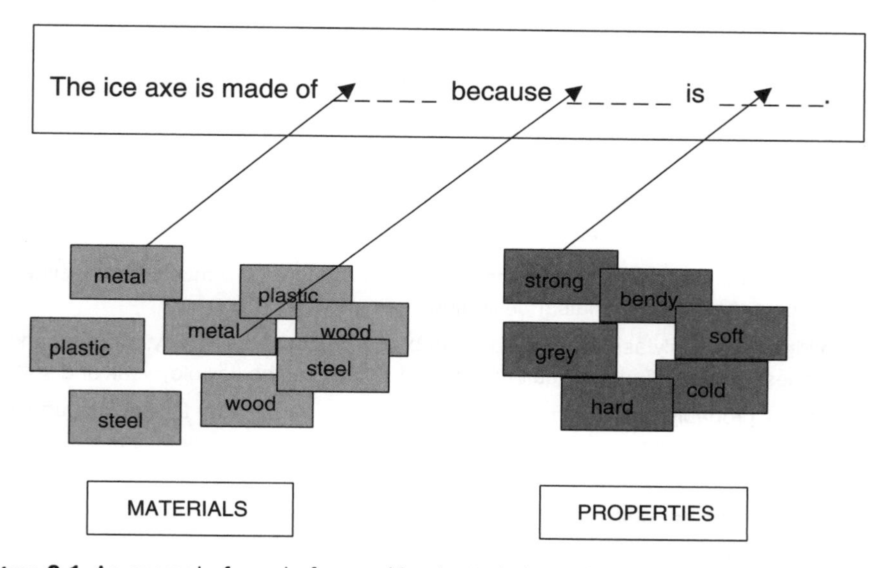

Figure 6.1 An example formula for word banks to help pupils distinguish between materials and their properties

the mask *is made of* keeps water out and is see-through, so you can see underwater'. For children with the ability to read, I often use pre-prepared word banks to help them to construct the appropriate sentences as shown in Figure 6.1.

It is also important to make sure you create opportunities that give emphasis to identifying similarities between materials, not just differences. At the same time you should ensure you check for confusions of terminology with the children and include materials to challenge their potential misconceptions. For example a pullover made of fleece will be soft but careful examination of the fabric will show that it is not smooth.

Conceptual step three – ensuring both solids and liquids are introduced

It is very easy when discussing materials with the children to focus too much on solids and so, right from the start other states of matter should also be introduced. Since gases cause all manner of problems, I begin by focusing on solids and liquids and only later, once the children have grasped the basic principles, do I introduce some common gases.

You might note that the concepts of solids, liquids and gases are often not mentioned specifically at key stage 1. It is true to say that complex models of solids, liquids and gases are beyond the youngest children but state of matter at room temperature is a property of materials that they will certainly notice and comment on. Key stage 1 children are also often required to explore changes in materials such as melting. In the light of these points I have found it extremely helpful to begin simple discussions of states of matter at key stage 1. Indeed, attempting to gloss over these issues when the children are themselves aware of them and developing their own ideas about them is only likely to lead to the establishment of misconceptions that early discussion could challenge and prevent from becoming embedded.

Be warned!

The different states of matter confuse children in many ways. Some common problems to look out for are listed following.

1. Since solids are the most concrete of materials children are much more inclined to accept them as materials than liquids or gases.
2. Children often only associate solids with properties such as rigidity, strength and hardness and can find it difficult to accept that they can be flexible, weak and soft.
3. Water is probably the liquid that most children use as a model for thinking about liquids. This means that they will often become confused by viscous liquids, e.g. honey, and may not think of them as liquids at all.
4. Connected to the above point, you will find that the children will be particularly confused by substances such as powders, emulsions, foams or pastes.

Challenging the children's ideas

The simplest way of addressing the above difficulties is to ensure the children get the chance to observe and discuss a wide range of solids and liquids, including difficult examples. To give an example from the dressing up context, when I use my alpine mountaineering equipment, I ensure that I have included examples such as:

- various liquids, such as warm water in a flask and liquid antiseptic in a first aid kit;
- soft, fleece fabrics to challenge the notion that solids are rigid and hard;
- difficult to classify substances, such as sun cream;
- powdered, instant tea to show that lots of little lumps of solid can flow around in a liquid-like fashion.

The discussion of examples like these will not only let you elicit the children's ideas effectively but will allow you to begin to challenge them. For example, if the children think the powdered tea might be a liquid because it can be easily poured, let them examine it with a hand lens and see that it is made up of lots of little separate lumps of solids.

Hint!

When encouraging children to observe and suggest differences between solids and liquids, I find the best feature to concentrate on at key stage 1 is 'pourability'. All the liquids that children of this age will be introduced can be poured, even if very slowly. (It is true that glass is often considered an extremely viscous liquid but this is a concept well beyond young children and will only cause confusion if you attempt to introduce it.) Remember though, that collections of lots of lumps of solids, such as sand, or talcum powder, can also be poured and so make sure you challenge the children's ideas regarding these, as outlined above.

Conceptual step four – extending the above to common gases

Gases are the most difficult of the materials for children to understand but, even at key stage 1, it is well worth beginning to introduce the children to thinking about them. I have almost always found that even very young children have encountered the term 'gas' and have ideas about gases, although these have typically included many misconceptions. Given this, I have found it better to begin discussing gases from an early stage. Young children need not have a complex understanding of what a gas is but it is helpful that the children: recognize gases as materials; understand that gases, just like other materials, have different properties; know that there are lots of different types of gases; and know a little about some common ones. This brief, early introduction can help alleviate more serious problems later on in the children's conceptual development.

Eliciting ideas

In the dressing up context it may be difficult to find a lead in to discussing gases. When I use my SCUBA gear it's easy to demonstrate air inflating buoyancy aids and being used for breathing. Likewise, if the children are meeting fire officers, they can discuss their breathing apparatus, but otherwise gases might seem rather obscure. However, one useful lead-in is provided by the wind, a phenomenon dependent on moving gases! Some of the clothes the children will be examining will be designed to keep the wind out. Make sure you ask the children questions, for example, probing what they think the wind is or what they think air is. It is also worth widening the discussion and asking what they think about gases generally. You are almost certain to find that you have a variety of misconceptions to challenge.

Be warned!

Children show many misconceptions in this area; some that you are likely to encounter are:

1 The concept of a gas is often limited to that of domestic gas with consequent views of all gases being dangerous, explosive or poisonous.

2 The children often do not regard gases as materials at all; hardly surprising given the seemingly incorporeal nature of the gases the children commonly encounter.

3 Terminology and discrimination will be limited. Some children will not regard air as a gas, thinking of it as simply 'air'. On the other hand, I have worked with other children who have seen the terms gas and air as synonymous.

4 Many children regard some non-gases such as smoke or flames as gases. Particularly common is the classification of sprays, mist, fog, clouds and steam as gases when they are, in fact made up of tiny drops of liquids suspended in the air.

Challenging the children's ideas

One of the first steps to take is to challenge the notion that 'gas' only refers to the domestic gas that the children may be familiar with as a fuel for heating and cooking. Such gases should be discussed but you should also be ready to introduce the children to other common gases. The most obvious is air and at this stage I don't worry about the fact that air is itself a mixture of other gases. Concentrate on attempting to establish that gases are materials just like solids and liquids and, like them, have properties. It is particularly important to challenge the notion that gases are non-corporeal. Accordingly you should highlight their properties and what 'they can do'; for example inflate balloons. Ask questions such as: Can you see air? Can you smell air? Can you feel air? etc. These questions provide a useful means of both eliciting the children's ideas and challenging them through discussion. The discussions will also lead into developing the correct terminology and to challenging incorrect ideas such as mist or steam being gases.

When you are encouraging the children to observe and suggest differences between gases and solids and liquids you will not have the benefit of one, more obvious, distinguishing feature such as 'pourability' that can be used to distinguish liquids from solids. Examples of some useful conceptual observations that I have encountered young children making have included:

- 'gases are very light';
- 'gases can easily spread out all around you';
- 'gases can be used to blow things up' (inflate them); and
- 'gases can make bubbles in liquids'.

If you can encourage your children to make observations such as these, they will be well on their way to a basic conceptual understanding of gases.

Hint!

The misconception that mist is a gas can be challenged by getting the children to look closely at it. This can be done using a simple plant mister. If the children look closely, they will see the mist is made up of tiny droplets, which can be reinforced by hanging threads in the mist cloud to catch the drops of liquid. The discussion can then be extended to fog and clouds.

Attempting a similar activity with steam obviously presents serious safety issues but hopefully the children will be able to extend the learning regarding mist to steam and accept that it too consists of tiny droplets of liquid.

Hint!

Another excellent gas to introduce the children to is carbon dioxide, the gas that makes the fizz in their drinks. Discussion can begin by examining the drinks. Lots of questions can be used such as: What do you think causes the fizz? What do you think the bubbles are? Where does the fizz go when the drinks get flat? A great idea is then to use a fizzy drinks maker to turn flat drinks into fizzy ones. This helps the children conceptually in several ways. By noting how the gas is stored and what it does, they learn more of its properties. The activity also helps show gas as a corporeal material that is stored in a cylinder and then added to the liquid.

Carrying out simple, fun activities like these with young children will greatly help their progress when they come to tackle more complex issues later.

Conceptual step five – sorting objects by both material and property and classifying them as solids, liquids or gases

Once the steps above have been completed, it is very straightforward for the children to begin to be able to sort objects by both material and by property. Indeed these sorts of tasks often make useful assessment exercises regarding the concepts outlined above. Do make sure that you get the children to sort objects at different stages by both material and by property. This will give you a much more secure insight as to how they have progressed. Remember also to include some of the tricky examples outlined to ensure the children's misconceptions have been addressed.

Conceptual step six – the uses of materials and linking these to their properties

Once again, having addressed the trickier conceptual challenges outlined above, this step now becomes quite straightforward and indeed, the example dressing up context will already have helped the children to think in these terms. A return to the specialist equipment is now a good idea with the children being asked why certain materials have been chosen for certain objects. Again, such tasks provide good assessment opportunities.

Hint!

This stage can support some great investigative work with the children being asked to test certain materials with respect to how suitable their properties are for particular uses. The learning context is improved if the children have a good reason for carrying out the test, for example they might need a waterproof groundsheet for a picnic expedition. Another context that I have found successful is equipping fictional characters for a variety of careers, or adventures. These characters can be anyone from the class teddy to cartoon characters. Figure 6.2 shows one such example.

On a trip to a local nature reserve, the children learnt that owls don't like the rain as their feathers are not waterproof. They then helped Powell the owl test a variety of materials to make him wet weather gear to keep him dry when stuck in his tree, unable to hunt because of the rain.

Figure 6.2 Testing how waterproof materials are

Conceptual step seven – where do materials come from?

This is a topic the children usually find interesting and it is nowhere near as conceptually difficult as some of the other aspects of materials. It is still important to pre-assess your children's ideas though; just think of the old story about the child who didn't realize that milk came from cows, believing instead that it, 'came from the supermarket'!

Be warned!

Many science schemes suggest drawing a distinction between natural and man-made materials but I have found this to be an unhelpful and confusing distinction for the children. Consider some of the potential sources of confusion. Children will usually easily see wood or wool as natural materials, after all they come directly from living organisms, but the forms in which the materials are encountered by children are typically heavily processed by humans. A material such as steel causes more problems. Metals are usually seen as natural but steel is only obtained by heavily processing iron ore and combining it with other substances, for example carbon. How many materials are purely man-made? Where should the children draw the line between natural and man-made? There are a wide variety of plastics and similar polymers that are synthesized by humans but what of the materials from which they are synthesized? Are they not natural? Some radioactive elements can only be created under laboratory conditions but these are too obscure for primary school situations.

Hint!

I find the best way to avoid the above problems is to scrap the distinction altogether. Instead I concentrate on getting the children to use questions such as: Where did the material come from? Or, how did it get to be the way we find it now? By concentrating on 'the story of where the materials come from', you will avoid some of the contradictions and confusions of imposing an overly narrow and rigid classification on the children's understanding.

7

Developing an initial understanding of changes in materials

Introduction

A glance at the conceptual steps listed for this chapter will show that there is an emphasis on the children's noticing and observing changes in materials rather than attempting to explain them. Even young children will be able to spot the causes of changes to materials but explaining why the materials change as they do is likely to be beyond the majority of key stage 1 children as it demands a simple particle model of matter. Accordingly, I have always found that it is best to emphasize exploration and observation in this topic at key stage 1 and to leave the explanations until the children have been introduced to the particle model of matter in key stage 2 (discussed in Chapter 8).

Where this topic fits in

The content outlined in this topic should typically be covered in key stage 1 and should be accessible to even the least experienced children. Learning about changes in materials is routinely treated as a different topic to classifying materials and their properties. This makes sense as the children will only have the necessary concepts to explore changes in materials after they are familiar with concepts of classification and property. However, it is important to realize that the two topics are closely linked and that the children will have to revisit their earlier work on materials and their properties when they come to explore changes in materials. The work covered in Chapter 6 is therefore essential to consider when helping children learn the concepts in this chapter.

Links to other topics

The close links to the classification of materials and their properties have already been mentioned. The concepts in this topic are also important, as the observations the children will make will provide the foundation for them when, in key stage 2, they begin to develop models to explain why materials change.

Another link that might seem surprising is to the topic of forces. One of the key concepts young children must learn about forces is that they can change the shape of objects. When the children are learning about changing materials by manipulating them, they will be changing their shape by applying forces to them and this makes for an ideal link between the two topics. When managing the science curriculum in schools, I have often planned to address the initial conceptual step of this topic (learning about changing materials by manipulating them) through an introductory topic on forces where the children have to explore how materials can be changed using pushes and pulls. This short forces topic can come between an initial topic on classifying materials and their properties and a further one on changing materials where the children explore more complex ways of changing materials such as heating and cooling. Details of how forces relate to changing the shape of materials can be found in Chapter 18.

A context for the topic

It is possible to cover the concepts of this topic through a series of different activities connected only by the fact that they all involve looking at changing materials but I find such a de-contextualized approach can be a bit dull for the children. It is not difficult to find a theme to link the activities and by far the most popular with the children I have worked with is a theme involving food! There are obvious health and safety issues when working with food technology in school but none that will prevent excellent learning, so long as they are carefully planned for and addressed.

As well as food frequently linking to the technology curricula that children follow, it often has other cross-curricular links. Over the years I have seen lots of successful food themes being used at key stage 1. Examples include: planning a teddy bears' picnic; preparing food for various festivals; preparing food linked to a place the children have learnt about through geography; preparing food linked to the children's history topic; and planning parties for characters the children have encountered through reading, or have themselves invented in their imaginative writing. Any of these contexts will support a range of simple, safe, food preparation activities that will allow the children to learn about changes in materials.

> ***Remember!***
>
> Such cooking topics are a fun and effective learning context but they do demand careful attention to the appropriate hygiene and health and safety procedures, although this in itself makes for valuable learning for the children.

Conceptual step one – noticing how materials can be changed by simple manipulation

This is actually conceptually very straightforward and makes a good link to, and revision of, the children's work on the properties of materials. It is so straightforward,

Table 7.1 Conceptual steps and example learning objectives for developing an initial understanding of changes in materials (all of these learning objectives are suitable for coverage at key stage 1)

Conceptual step	*Example learning objectives*
	Children will learn:
Noticing how materials can be changed by simple manipulation.	how the shapes of common materials can be changed by manipulation; that when the above happens, the material has not changed into a different substance; to link their observations to their understanding of the classification of materials and their properties;
Noticing how materials can be changed by heating, cooling and being combined with other materials.	how common materials can be changed by heating, cooling and adding to other materials; to understand that when the above happens, the material can sometimes be returned to its previous state and sometimes not; to link their observations to their understanding of the classification of materials and their properties.

that I tend to elicit the children's ideas on these concepts at the same time as pre-assessing their ideas about how materials can be changed in more complex ways, for example by heating.

Eliciting ideas

In a cooking themed example, the easiest way in which to find out the children's ideas on changing materials by simple manipulation is to set them various tasks that will involve changing materials. For example, if the children are planning An Easter Teddy Bears' Picnic, they might make Easter nests from cornflakes covered in melted chocolate. Before beginning to cook, examine the ingredients with the children. It's a good idea to first revise earlier work by getting the children to identify the materials and to discuss their properties. Then ask the children how they think the materials might be changed during making the nests. Ask them what other ways they think the materials could be changed. The discussion can cover not just simple manipulation, such as breaking off a square of chocolate, but also the more complex changes, such as melting the chocolate, which are addressed in the next conceptual step. When eliciting the children's ideas through such discussion, it is important to encourage them to link their predictions to what they know about the properties of the materials under consideration.

Be warned!

You are unlikely to encounter misconceptions directly related to this area. Sometimes the children may be mistaken factually, for example suggesting that a spoon is too stiff to bend, but such errors usually represent a simple lack of experience rather than any serious conceptual problems. If you do find misconceptions they are more likely to relate to such issues as distinguishing materials from their properties and advice on challenging such misconceptions is given in Chapter 6. Sometimes however, children may think that changing a material, even by simple manipulation, turns it into a different material. For example, I remember a child telling me that crushing cornflakes, 'turned them into cornflake dust' and that this was, 'different stuff', by which the child meant a new material.

Challenging the children's ideas

As mentioned above, you are unlikely to have serious misconceptions to address here. The secret is to have lots of different cooking activities that involve handling a wide range of materials with many different properties. Throughout the process, ensure there is lots of discussion about how the materials are changing and relate this to the properties of the materials. Be ready to help the children with lots of questions to aid their thinking. For example in the powdered cornflake example above, it was important to focus the child's attention on the fact that the dust was still cornflakes, just in very small pieces. I asked questions such as: did you add anything to the cornflakes? Or, did you take anything away from them? And, do you think they would still taste like cornflakes? Analogies may also be very helpful. In the example above, I asked the child to think about cutting a cake up into pieces; did that make it something different from cake? The child then accepted that the dust was just very small pieces of cornflakes.

Conceptual step two – noticing how materials can be changed by heating, cooling and being combined with other materials

Again, the cooking context is ideal for developing these concepts and, once again, the children should be encouraged to link their observations and discussions to the properties of the materials they are working with.

Eliciting ideas

As discussed above, this can be done by discussing the recipes the children will be working with and asking them do they think the materials will change and how. The secret here is to ensure that the cooking activities illustrate a wide variety of changes in materials. Not all the changes have to be discussed at once but it is important to plan carefully to ensure the children encounter an appropriate range of changes. The following list illustrates the examples I aim to cover with the children.

1 Heating leading to simple changes of state e.g. melting chocolate.
2 Cooling leading to simple changes of state e.g. freezing lolly mixture. (N.B. heating and cooling are not quite as straightforward as they appear. Some useful ideas about the more complex issues involved can be found in Chapter 10.)

3 Heating leading to more complex, irreversible changes e.g. baking biscuits.
4 Burning! Obviously care is needed here but the children often discuss burning as something to avoid when, for example, baking.
5 Simple instances of mixing materials e.g. mixing icing sugar and 'hundreds and thousands' to sprinkle on top of a cake.
6 The particular mixing example of *dissolving* e.g. using instant lemon tea to make an iced tea.
7 Instances of mixing materials that result in their changing in irreversible ways e.g. adding water to cornflour to make a thickening agent for a curry.

It is also important to ensure that the examples involve not just solids and liquids but also gases and to link this coverage to the work discussed in Chapter 6. Gases are tricky but a great way to get the children to think about gases is to use a fizzy drinks maker. These are a simple and fun way to introduce the children to thinking about gases in a context they are familiar with. And the drinks maker also fits well with a cooking theme.

A cooking theme also offers another opportunity to get the children thinking and learning about gases; that is, through their sense of smell. Hopefully your cooking activities will produce a range of delicious aromas and this allows you to elicit the children's ideas as to what a smell actually is.

Be warned!

There are a number of common misconceptions that it is wise to be prepared to deal with:

1 Many young children think materials have always existed in their current form and do not appreciate that they have often been heavily processed. It may be hard to believe, but I remember asking some children where wood comes from and one replied, 'you get it in *B&Q*'.
2 Even when the children know that changes to materials occur during manufacture and processing, they may have factual confusions about the processes involved, or the raw materials. For example, I recall a child who initially believed that all fabrics, 'came from sheep'.
3 Children often have difficulties regarding changes of state (solid, liquid and gas) of materials. I find they rapidly appreciate the change of state from solid ice to liquid water but are confused by the concept of water as a gas (water vapour.) Also, they often find it difficult to transfer these concepts to other materials, for example I have found many children initially refuse to believe that vinegar could exist as a solid.
4 You should also be aware that misconceptions often arise from the children confusing melting and dissolving; and also confusing heating and burning.

It is also interesting to note how misconceptions can change along with social and cultural changes: I have now met several children who believed that the only source of metals was recycling!

Challenging the children's ideas

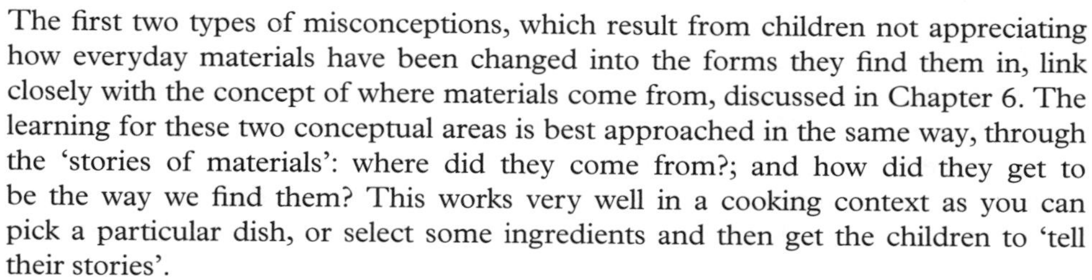

The first two types of misconceptions, which result from children not appreciating how everyday materials have been changed into the forms they find them in, link closely with the concept of where materials come from, discussed in Chapter 6. The learning for these two conceptual areas is best approached in the same way, through the 'stories of materials': where did they come from?; and how did they get to be the way we find them? This works very well in a cooking context as you can pick a particular dish, or select some ingredients and then get the children to 'tell their stories'.

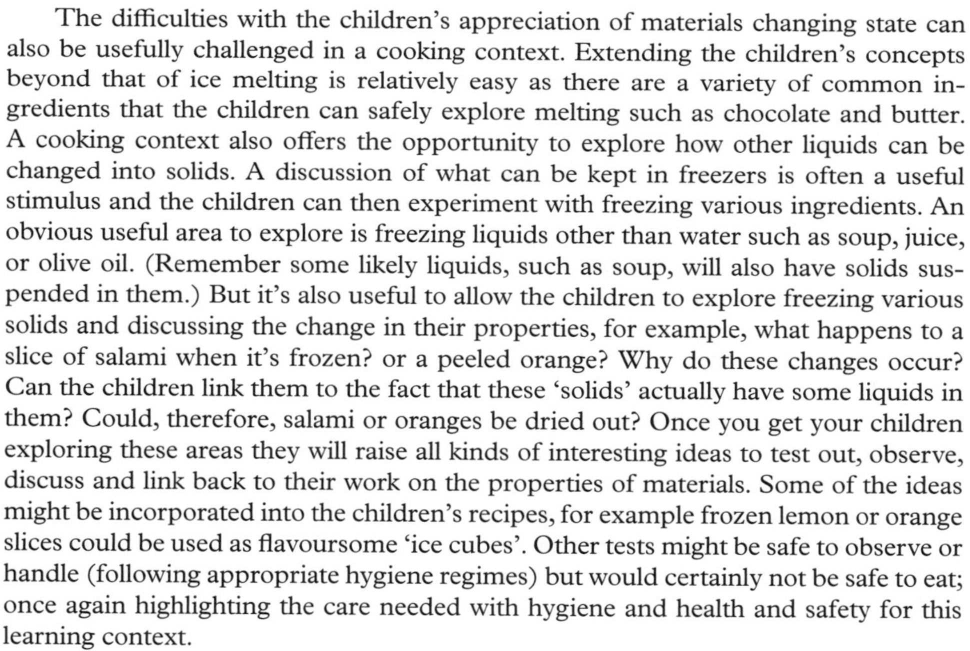

The difficulties with the children's appreciation of materials changing state can also be usefully challenged in a cooking context. Extending the children's concepts beyond that of ice melting is relatively easy as there are a variety of common ingredients that the children can safely explore melting such as chocolate and butter. A cooking context also offers the opportunity to explore how other liquids can be changed into solids. A discussion of what can be kept in freezers is often a useful stimulus and the children can then experiment with freezing various ingredients. An obvious useful area to explore is freezing liquids other than water such as soup, juice, or olive oil. (Remember some likely liquids, such as soup, will also have solids suspended in them.) But it's also useful to allow the children to explore freezing various solids and discussing the change in their properties, for example, what happens to a slice of salami when it's frozen? or a peeled orange? Why do these changes occur? Can the children link them to the fact that these 'solids' actually have some liquids in them? Could, therefore, salami or oranges be dried out? Once you get your children exploring these areas they will raise all kinds of interesting ideas to test out, observe, discuss and link back to their work on the properties of materials. Some of the ideas might be incorporated into the children's recipes, for example frozen lemon or orange slices could be used as flavoursome 'ice cubes'. Other tests might be safe to observe or handle (following appropriate hygiene regimes) but would certainly not be safe to eat; once again highlighting the care needed with hygiene and health and safety for this learning context.

Appreciating changes of state to gaseous states is more complex and it is little wonder that the children often have misconceptions in this area. However those savoury aromas can be very useful! Getting the children to think about how we smell can help them learn more about gases and allow them to take the first steps in thinking about evaporation. For example, I use strategies such as leaving out plastic pots with vinegar in them. The smell of the vinegar spreads through the room and this allows discussion of what smells actually are. Helping the children appreciate that they can smell the vinegar because vinegar vapour (gas) is spreading through the room and mixing with the air can be a difficult thing to do. Undoubtedly, the best way to do this is to introduce the children to a simple particle model of matter but this is likely to be rather beyond many key stage 1 children. The use of such models is discussed in

Chapter 8 and, if you are working with capable, older key stage 1 children you might like to consider introducing them to the model. Even if you feel a particle model might be difficult for your children to grasp, it is important to challenge the likely misconception that the 'smell' is something non-corporeal and not the result of a physical material. I always ask the children how they think they can smell vinegar even when they are some distance from the bottle. I've found it useful to start the thinking with questions such as, how do you smell things? The children easily appreciate that they use their noses and you can then ask questions such as, how does the smell get to your nose? Initially, the children's ideas tend to have little to do with gases. Typically for young children, they often give a response that involves circular reasoning, rather than any kind of explanatory model. The most common response is usually a variation on, 'I can smell it because the smell goes through the room'. You can then ask the children what they think the smell actually is and how it, 'goes through the room'. Interestingly, when I use such questioning sequences, I find that it seems to help some children appreciate that the smell must be 'something' and I regularly get suggestions such as, 'the bits of smell go through the air', or, 'the smell is like the wind, it sort of blows through the room'. Such ideas are moving towards the concept that some vinegar particles escape from the liquid as a gas and drift through the room. Undoubtedly some younger, or less capable children will not come up with such ideas themselves but discussion with yourself and other children can help them appreciate what's going on. Lots of talk about the various examples the children have encountered that show them that gases are actually materials (see Chapter 6) will help them begin to appreciate that little bits of the vinegar turn into a gas and drift through the air, and that our noses detect these little bits and that's what the smell is.

Although a full exploration of evaporation is more appropriate to key stage 2 work, if gases are being discussed in this way at key stage 1, the topic is likely to be touched upon. You may find the discussion of evaporation in Chapter 8 helpful even if you are not fully exploring this topic with your key stage 1 children. Consideration of these concepts can often help challenge misconceptions at an early stage. For example, a very common misconception is that, for evaporation to occur, a heat source is required such as a radiator, the Sun or a cooker element. By using strategies such as the discussion of the smell of vinegar spreading through the room, the children will be introduced from an early stage to the fact that liquids can turn into gases without such heat sources.

8

Developing a more advanced understanding of classifying materials and their properties

Introduction

Just as was mentioned in Chapter 6, the concepts introduced to children through the topic of materials and their properties, particularly if a simple particle model of matter is used, will stand them in good stead when it comes to explaining what they encounter in other areas of science.

Where this topic fits in

The concepts covered in this section are all likely to be covered in key stage 2. However, they are not particularly complex and I have found they can be covered early in the key stage. There is an extra advantage in doing this, as early coverage of these concepts will help the children when they come to tackle other, more complex topics in other areas of their science learning, such as light and electricity.

The concepts dealt with in this section build heavily on those introduced in Chapter 6, and it is essential that your learners are secure in the concepts introduced in that section before proceeding to those outlined here. Indeed it is likely that many of the conceptual steps and likely learning objectives explored in that chapter will have to be revisited in key stage 2, albeit in more complex contexts, to secure the children's understanding.

Links to other topics

This topic builds heavily on the concepts of Chapter 6, with the main difference being that the children are now introduced to a simple particle model to help them explain their observations. This particle model will also be very useful to the children in making sense of the discoveries they will make when learning about: changes in materials (Chapter 9); thermal insulation (Chapter 10); magnetism (Chapter 11); and electricity (Chapter 16).

A context for the topic

The concepts of this section can be tackled by a variety of imaginative topics.

For older key stage 2 children, I have often used contexts that build on grouping and classifying materials and involve the children in testing the properties of materials for a variety of uses. Some of the most successful contexts have involved survival situations. For example, once when using Robinson Crusoe as a stimulus for imaginative writing, I set up a shipwreck situation where the children imagined they had been wrecked on a desert island. The children were given access to a range of salvaged and naturally available materials and had to decide how they would use them to survive. The children then conducted tests on the materials to see if their suggestions would have helped them stay alive. For example, some children had suggested making huts to stay dry and carried out an investigation comparing various materials to see which was the most waterproof. I have also used a 'space wreck' variation of this context in conjunction with a science fiction theme in imaginative writing. This worked very well too, with the added advantage that a larger range of modern, more heavily processed materials could be added.

Conceptual step one – extending previous work on the properties of materials with reference to a simple particle model of matter

Any of the above topics will allow the children to revisit the learning objectives outlined for classifying materials and their properties in the last topic. In many respects the conceptual aspects will be the same but I have found that there are four main areas where I expect to see progression in the children's understanding:

1 Breadth of knowledge of materials. For example, I expect the children to be able to recognize a greater number of materials. This will also involve a greater factual acuity, for example whereas key stage 1 children might just talk of 'metals', more experienced children should be able to differentiate between a range of common metals, such as copper, iron, brass etc.

Table 8.1 Conceptual steps and example learning objectives for developing a more advanced understanding of classifying materials and their properties (all these learning objectives are suitable for use at key stage 2)

Conceptual step	*Example learning objectives*
Extending previous work on the properties of materials with reference to a simple particle model of matter.	Children will: learn a simple particle model to explain the properties of solids, liquids and gases; be able to explain properties of materials with reference to a simple particle model.

2 Breadth of knowledge of properties. I expect the children to become aware of a wider range of properties, often linking this with their work on changing materials. Examples of some extra properties I would expect the children to consider are properties such as solubility in water (here there are links to Chapter 9) or whether a material melts or burns (or potentially both) when heated.
3 Technical terminology. For instance 'bendy' should become 'flexible' and 'see through' become 'transparent'.
4 Explanations of all the above in terms of a simple particle model of matter.

The first three aspects are all straightforward and can be tackled using exactly the same sorts of strategies that were outlined in Chapter 6. The fourth aspect is more difficult and demands some careful thought. I begin by getting the children to consider solids, liquids and gases.

Eliciting ideas

By this stage, the children should have been introduced to a range of materials in the basic three states of matter: solid; liquid; and gas. As outlined in Chapter 6, they should also have begun to make observations about the different states, for example regarding the 'pourability' of liquids. However they will not have begun to consider why, for example, copper is a solid at room temperature compared to water that, at the same temperature, is a liquid. It is now interesting to probe their ideas regarding these matters and the easiest way to do this is to simply ask them what they think.

Be warned!

Ideally you will be addressing such concepts fairly early in key stage 2 and in my experience such children are often pretty baffled by these ideas. If you have created an atmosphere where ideas are valued, you will find that some children offer various imaginative suggestion but these are likely to be wide of the mark. Often they can't get beyond the actual properties to give reasons for the state of matter and so may say things such as, 'the copper is solid because it can't flow around'. Equally, following on from key stage 1 work on changing materials, they often focus on temperature as an explanation for a material's state. Accordingly, you may hear statements such as, 'the copper is a solid because it's cold and the water is warmer so it's a liquid'.

Challenging the children's ideas

Misconceptions such as the above can be challenged fairly simply. The first can be addressed simply by emphasizing that it is a reason that we are looking for. So a suggestion such as, 'the water's a liquid because it can be poured', can be countered with a response such as, yes, but what makes liquids pourable? You can imagine the possibly circular reply, 'because they're runny!' Even so, with patience, it is possible to get the children to focus on the issue of explanation. I have often found that some practical observations and discussions help, for example looking at ice and water and asking why the water can flow and the ice can't. This usually raises the issue of temperature and the children are likely to focus on the fact that the ice is cold. This is fine so long as it doesn't lead to the other misconception listed, that of believing that solids are always colder than liquids. However, even if this problem is encountered, it is relatively easy to challenge simply by measuring the temperature of some solids and liquids; if left to sit in the classroom, a bowl of water and a piece of copper will both be at room temperature. It is also useful to encourage comparisons such as the fact that ice will melt even in a cold room, whilst copper needs a very hot furnace to melt it.

Even so, in my experience, many children find it hard to come up with ideas to explain the changes and, in this example, you may find that you are not so much challenging ideas, as introducing the children to a model for their thinking.

Introducing a simple particle model to explain states of matter

In fact, the good news is that, if you introduce the model carefully, I have found that the children usually accept it and can quite happily use it as an explanatory tool.

I begin by suggesting to the children that all materials are made of 'bits' and that these 'bits' are the material. Thus copper is made up of copper 'bits', wood is made up of wood 'bits' and water is made up of water 'bits'. It is not difficult for the children to accept that a special word for these bits is *particle*. At this stage I do not worry about attempting to introduce the concept of what chemists would term elements and compounds. (Elements are materials made up of just one type of atom and compounds are made up of different sorts of atoms chemically joined together. For example copper is an element, being made up just of copper atoms. Water, on the other hand is a compound made up of hydrogen and oxygen particles joined together, as well as, in most cases, having other substances dissolved in it. Wood will be even more complex and will consist of combinations of various compounds.) However, when these concepts come to be introduced later, when the children encounter chemical reactions in their work on changing materials, I have found the children assimilate it without much difficulty and, in fact, some children may have already heard of such examples. I often find many even quite young children have encountered water as H_2O and know it has hydrogen and oxygen in it (although they are often insecure in understanding that the water *is* the hydrogen and oxygen). It is also helpful to not just talk of water being made up of particles, wood being made of particles etc. but to stress that water *is* lots (understatement!) of water particles and that wood *is* lots of wood particles. This helps challenge the common misconception that children have of, even when they accept the existence of particles, believing that the particles are something in addition to the water, wood, or whatever.

The key fact at this stage is that what determines the state of matter of a material is how these particles are joined together. In a solid they are joined relatively rigidly and cannot move around each other; in a liquid, the particles are joined together but can easily move around each other, allowing the liquid to flow; in a gas, the particles are not joined to each other and can move around independently.

Hint!

By far and away the best way for the children to grasp the model is for them to act as particles in the various states of matter. In such cases each child represents a particle and they use their arms to represent the bonds between particles.

When acting out a solid, the children should stand with their arms rigidly linked. Be careful here! The children tend to instinctively bunch up closely when acting a solid and, since most solids take up less volume than their liquid equivalent, this may be fine. However, water, the material that is most safely and easily accessible to the children as solid, liquid and gas, is an exception. Water takes up a greater volume as ice than as liquid water, so, if the children are acting as ice, they should extend their arms, but keep them rigidly locked together.

The children must link arms firmly but without squeezing up, as water is unusual in that it expands in solid form. The children should remain standing in the one spot.

Figure 8.1 Diagram of children acting out water in its solid form as ice

When acting out a liquid, the children still need to touch each other but can now move around each other. I find it helpful to give them the rule that they should always keep at least one hand on another child's shoulder.

Here the children can move around but they must always remain in contact with at least one other child. I usually give a rule that they must keep one hand on the shoulders or arm of another child but that they can move around within the 'drop of water'.

Figure 8.2 Diagram of children acting out water in its liquid form

When acting out a gas, they can move around without touching each other.

Now the children put their arms down and, although they might sometimes bump into another 'particle', they move independently through the space available to them.

Figure 8.3 Diagram of children acting out water in its gaseous form as water vapour

Using this simple particle model to explain other characteristics of materials

As well as states of matter, this simple particle model can now be used by the children to explain other properties of materials that they will have encountered, described and classified materials by.

For example, in very hard materials, the particles must be difficult to scrape off each other. Children will now usually quickly appreciate that this must be because they are joined very tightly together. For a fragile, easily broken material, the particles must be joined less tightly together. For materials that can be stretched, it must be possible to stretch the joins between the particles.

Once again, the best way to help the children appreciate these sorts of concepts is to let them act them out with the children themselves being the particles and using their arms to join themselves together. (An example of this relevant to forces is discussed in Chapter 19.)

What do children really understand by such models?

There is no doubt in my mind that, when using such models, the majority of children only have the vaguest ideas about what exactly these particles are and how materials are constructed out of them. Typically, for instance, children do not comprehend the incredibly small size of the particles, nor the astronomically huge numbers involved. (But then who amongst us really does?) I have never found such matters a problem. The children can happily adopt and use the model to predict and explain common phenomena they encounter and they can do so without reinforcing or introducing pernicious misconceptions. If we attempted to wait until our learners could have a complete comprehension of the phenomena we introduce them to, we would be out of a job! Even quantum physicists admit they have no real idea what is actually going on at the atomic and sub-atomic levels they study, yet they too successfully use models to predict and explain the phenomena of the universe to amazing degrees of accuracy; there is no harm in the children doing exactly the same at their level.

9

Developing a more advanced understanding of changes in materials

Introduction

Chapters 6 and 8 discussed how to introduce and extend children's knowledge of classifying materials and their properties. This included introducing more experienced children to a simple particle model that they could use to explain such things as the different states of matter. Chapter 7 discussed how younger primary school children might be introduced to observing and discussing simple changes in materials brought about by manipulation, heating, cooling and mixing. At that basic level, much of the emphasis is on encouraging the children to observe changes and to build up a suitable vocabulary to discuss what they see. Once the children are more experienced, it is important to revisit changes in materials. This will involve introducing more complex changes, in the form of chemical reactions, and also helping the children construct models that allow them to explain the various changes in materials they encounter.

Where this topic fits in

All of the conceptual steps listed in this topic are best tackled in key stage 2. Research suggests that primary school children typically have difficulty with these concepts and certainly the most difficult ones, such as the chemical reactions involved in burning, or rusting, might be better left until upper key stage 2. However, I have found that, if children have been progressively introduced to the sorts of models previously outlined in Chapter 8, even younger children can employ simple particle models to explain phenomena in a scientifically appropriate fashion.

Links to other topics

This topic obviously follows on from the work on materials introduced in Chapters 6, 7 and 8. In particular, it builds upon the simple particle model of matter introduced in Chapter 8. However, the concepts discussed here are vital if the children are to understand, not just materials, but many other areas of science. For example, there are links to the concepts of thermal insulation and magnetism, discussed in Chapters 10 and 11, respectively. In addition some concepts will be essential for the children's understanding of light (Chapters 12 and 13) and electricity (Chapters 16 and 17). The concepts are even required if the children are to be able to explain some of the phenomena they will encounter in their work on forces (Chapters 18 and 19).

A context for the topic

One of the most successful contexts I have used for this topic is that of detectives or forensic scientists who have to use chemical techniques to gather evidence and solve mysteries. For example, the children might be asked to match a substance recovered from a crime scene with substances recovered from the homes of various suspects. No dangerous or difficult to obtain chemicals are required for such work; even everyday household materials such as cornflour, icing sugar, scentless talcum powder, or health salts can be used to set children educational and entertaining challenges. This sort of approach is also easy to differentiate. The difficulty of the task can be matched to the skills of the children. It is also easy to add extra layers to the tasks. For example, perhaps a sample has been recovered from the ground and is mixed with sand, or gravel: can the children purify the sample before testing it further? Or, what if a suspect may have hidden a substance by dissolving it in a liquid: can the children check for this and recover a sample?

Because of the ubiquity of materials in our daily lives, you may find that many cross-curricular links offer possible contexts for learning about changes in materials. For example, I was recently working with schools on a local history project that focused on prehistoric times. This offered many cross-curricular opportunities, including a fabulous opportunity to do work on materials and their properties. Working with local bush craft experts, the children examined the properties of naturally occurring materials, and how prehistoric peoples made use of these to produce an astonishing range of items from fishing line made from nettle fibres to temporary shelters made from hazel poles and thatch. The theme also allowed the children to focus on changes in materials, for example: they learnt about natural methods of fire making and examined the chemical reactions involved in burning and cooking; they also looked at making pottery and the chemical reactions involved in this. Considering the work of archaeologists and how different materials preserve, or not, offered more interesting opportunities to look at chemical reactions by examining questions such as: What rots? What conditions sometimes stop rotting? Why do bones dissolve in certain soils? Or, what happens to different metals when they are buried in the ground?

Conceptual step one – applying a simple particle model to explain mechanical changes in materials

This conceptual step is essentially a revision of concepts introduced in Chapters 6 and 7 with the particle model introduced in Chapter 8 then being extended to allow the children to explain their observations.

Table 9.1 Conceptual steps and example learning objectives for developing a more advanced understanding of changes in materials (all of these conceptual steps are most likely to be best addressed in key stage 2)

Conceptual step	*Example learning objectives*
	Children will:
Applying a simple particle model to explain mechanical changes in materials.	understand a simple model of atoms and molecules that they can use to explain changes in materials; learn to explain the effects of manipulating common materials in terms of a simple particle theory;
Applying a simple particle model to explain dissolving.	learn to explain dissolving in terms of a simple particle theory;
Applying a simple particle model to explain evaporation and condensation.	learn to explain evaporation and condensation in terms of a simple particle theory;
Applying a simple particle model to understand chemical reactions.	learn to explain basic chemical reactions in terms of a simple particle theory (e.g. water being added to health salts, burning and rusting).

Eliciting ideas

Either of the contexts outlined above provide many opportunities to elicit the children's ideas as to why materials change and behave in certain ways.

For example, in the prehistory topic example, you might ask the children what it is that makes wood springy, so that, when a drawn bow is released, it springs back to its original shape. Or, you might ask them what it is about wet clay that makes it easy to shape and mould.

Be warned!

Traditionally research has shown that children have very confused ideas regarding the fact that materials are made of particles. Typical problems include:

- believing that a material is simply 'the material' and seeing no need for any particulate model to explain this;
- even if the concept of particles is accepted, children may still persist in believing that the particles are somehow separate from the material;
- finding it difficult to believe, especially for solids and liquids, that there are spaces between the particles;
- having little comprehension of just how small the particles are and how many of them there are.

In turn, these sorts of problems often lead to children seeing no need to evoke any sort of model to explain the phenomena they encounter. For example, children will often advance a circular argument to explain why wood is springy such as stating, 'it is springy because it is flexible'. Of course, not all flexible materials are springy and, even if they were, this statement has not given an explanation for what causes springiness. The problems with such reasoning have already been discussed in the previous chapters on materials and Chapter 8 has already made some suggestions as to how to counter them. Once the children have reached the level of the concepts outlined in this chapter there is scope for helping them further and the good news is that research, and my own experience, show that, if the children receive some direct teaching about a particle model, that many of the above problems can be much reduced.

Challenging the children's ideas

The first step therefore in challenging the misconceptions is to provide the children with an alternative model to explain their observations.

To some degree this will already have been done if the approach outlined in Chapter 8 has been followed. This will have introduced the children to the concept that materials are made up of particles and that the particles are the materials and that this can be used to explain changes in state. As outlined in that chapter, I have also found that matters are further helped if the practice of referring to the particles by the name of the material is followed, for example: water (including when ice or water vapour) is made of water particles; copper is made of copper particles; and wood is made of wood particles. This is obviously a chemical simplification but it helps reinforce the fact that the particles are the material. However, by this stage of the children's progression, when more complex phenomena must be explained, it is appropriate to provide them with a more complex particle model. It is always best to have the children learn by investigation and discovery but sometimes there is no realistic alternative to some direct teaching. This is certainly the case when it comes to providing this model but when following the sort of progression outlined in the preceding chapters, I have never found problems with this: the model is not to give children 'the answers'; it is to provide them with a tool they can use to explain their observations.

Now it is necessary to go somewhat beyond the simple notion of materials consisting of particles and to talk more of exactly what these particles are. I usually begin by introducing the children to a simplified model of the atom (see Figure 9.1) and discussing how, for our purposes at primary school level, it is the most important sort of particle.

Of course, atoms are in reality more complex than this: the nucleus typically contains protons and neutrons; which are themselves composed of yet smaller particles. However, these details are not necessary for primary school purposes.

It is useful to emphasize several key points:

1 Atoms are incredibly small.
2 They are also incredibly numerous.
3 The nucleus is relatively big and the electrons are relatively small.
4 The positive and negative forces of the nucleus and the electrons tend to balance (the amounts of positive and negative usually equal out).

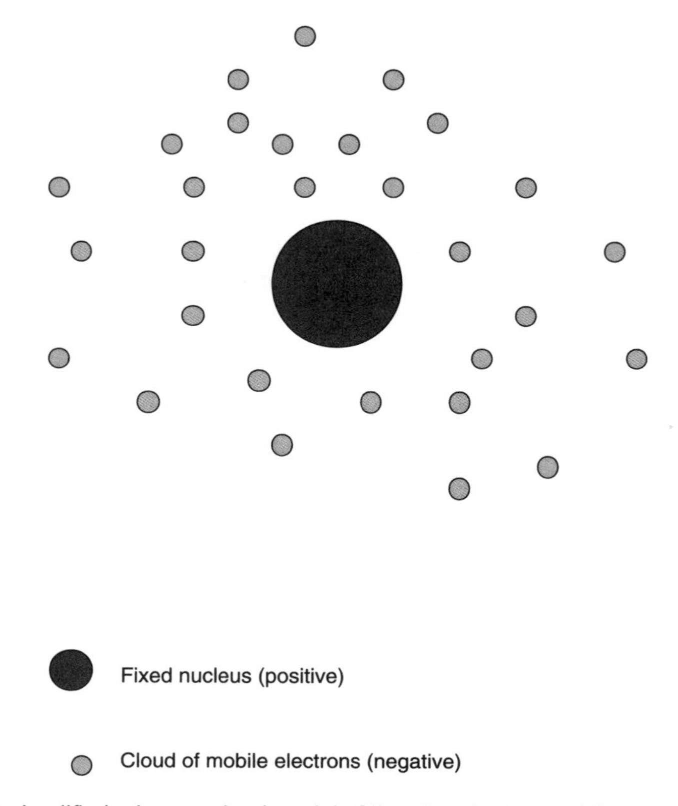

Figure 9.1 A simplified primary school model of the atom (not to scale)

5 Positive and negative attract one another but two negatives, or two positives repel each other.
6 Atoms stick together because of the attractions of their positive and negative forces. (It is not necessary at primary level to go into the complexities of bonding.)
7 Because of the tiny electrons hovering in their cloud, there is actually, relatively speaking, quite a lot of empty space between atoms, even when their positive and minus forces are sticking them together.

At this point, it is also worth discussing how some materials, for example copper or oxygen, are made of only one type of atom and are called *elements*; whilst other materials, such as water or steel, are made up of different sorts of atoms joined together and are called *compounds*.

The children can now use this model to explain why materials can be manipulated in certain ways. For example the wooden bow is springy because the way its atoms are joined together means that the space between some atoms can be stretched (those on the outside of the bow) whilst the space between other atoms is compressed (those on the inside of the bow). Then when the bow is released, the pushes and pulls between the atoms spring it back to its original shape. The wet clay can be moulded because the atoms must be able to move over each other. They can even be stretched a little before the joins between them break apart but they can't be compressed with anything like the springiness of the wooden bow.

To a large degree this model has only refined the concepts introduced in Chapter 8 but it can now also be used to tackle more complex phenomena.

Conceptual step two – applying a simple particle model to explain dissolving

Eliciting ideas

Once again, the example contexts give plenty of opportunity for the children to state their ideas. For example if, in the role of forensic scientists, the children have been testing various substances, they may discover that some dissolve in water, whilst others don't. It is very straightforward then to ask them what their explanation for this is.

Be warned!

Very young children, especially when colourless solutes such as sugar are used, sometimes state that the material 'disappears'. So long as proper health and safety is adhered to, this is easy to challenge as the sugar can be tasted in the water. Indeed most children readily accept that solutes don't just vanish. However this does not mean that they have a scientific understanding of what happens. Sometimes children have told me that the sugar 'mixes with the water'. Sometimes they have even told me that, 'the sugar becomes a liquid'. Both these statements are correct but often the children's understanding of how this occurs is incorrect. Frequently the explanation the children advance for this is that the water melts it. This is incorrect, though it is easy to see why it seems a logical possibility to the children, especially if they are encouraged to observe that materials dissolve more quickly at higher temperatures. Another common misconception that I have encountered is that solubility has something to do with the hardness of the material. For example, children have often told me that sand does not dissolve in water because it is hard but that sugar does because it is soft. Once again, there is logic behind this suggestion, even if it is not the case.

Challenging the children's ideas

Once children have got to grips with the model of atoms outlined above, it becomes much easier for them to appreciate what happens: that is, that the joins between the particles of the sugar break down so that it becomes so finely mixed with the water as to be part of a liquid solution. This can be demonstrated by a safe taste test, as discussed above, and by evaporating the water off to leave the sugar.

I have also found it helps the children greatly to act out the process using similar techniques to those discussed in Chapter 8. For example some children can be sugar particles and join themselves to make granules as in Figure 9.2.

Other children can then act as water and pour onto the sugar, dissolving it as in Figure 9.3.

Some children link arms and stand close together, acting as sugar particles clumped in granules. Other children stand by, acting as water. The water particles must always stay in contact with each other, with one outstretched arm touching another child's shoulders, but they can move around one another. (Not to scale.)

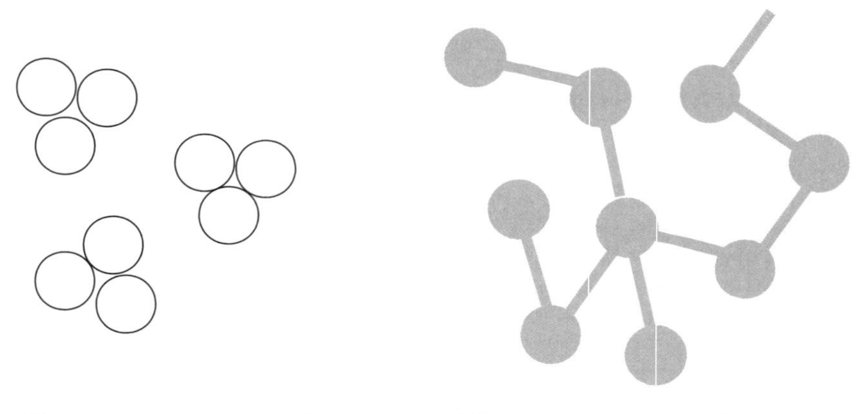

Children acting as sugar particles | Children acting as water particles

Figure 9.2 Diagram of children acting out sugar and water particles prior to dissolving

The 'water' is poured on and the sugar 'granules' dissolve, the sugar particles becoming part of the liquid solution. (Not to scale.)

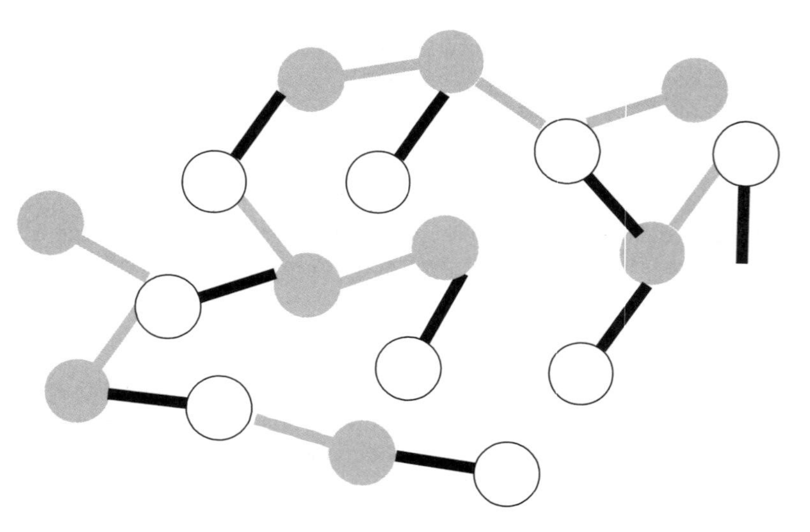

The sugar particles have now dissolved and become a liquid as part of the sugar solution.

Figure 9.3 Diagram of children acting out sugar particles dissolved in water

When such approaches are used, I have found that the majority of the children easily connect with the correct scientific ideas and can easily apply the model to act out and distinguish between dissolving and examples where small particles of a material do not dissolve and are simply held in suspension in a liquid.

Conceptual step three – applying a simple particle model to understand evaporation and condensation

Even at key stage 1, children should have been introduced to materials as solids, liquids and gases and a deeper understanding of states of matter is an important part of older children's work on classifying materials, as discussed in Chapter 8. So, to some extent, the children will already have been considering concepts relevant to evaporation and condensation. Now that they have been introduced to the extended atomic model, the children will be able to consider evaporation and condensation in more detail and explain it.

Eliciting ideas

Again, it is easy to see how the contexts outlined provide plenty of opportunity for the children to express their ideas: detectives might encounter invisible ink; forensic scientists might have to recover the solvent from an adulterated solution by distilling it; or prehistoric people might have to dry meat to preserve it. All of these examples would give excellent opportunities for you to elicit the children's ideas about evaporation or condensation.

Be warned!

Children typically have many misconceptions regarding evaporation and condensation.

All but the youngest and least experienced children are likely to accept that an evaporated liquid has not just disappeared but, just as with dissolving, they are likely to have quite sophisticated alternative (and scientifically incorrect) frameworks to explain what happens. For example, children frequently explain puddles evaporating by suggesting the water has soaked into the ground. In a similar fashion, I have heard children suggest that prehistoric people's animal skins dried because 'the water soaked into the fur'. A key difficulty here is that water vapour (water in its gaseous form) is invisible, so it is little wonder that children seek other, to them, more logical alternatives to the fact that the water has become part of the air around them.

Another very common misconception regarding evaporation is that extra heat is needed for it to happen. I have often heard very able children tell me that they can recover a solute from a solution but that they must heat it up to do this by, as one comprehensively thinking child told me, 'putting it on the radiator, leaving it in the Sun, or sticking it on the cooker'. It is true that heating will speed up evaporation (as will wind, or other air movements) but it is not necessary for evaporation to take place.

Condensation also causes problems.

The major problems stem from the fact, already mentioned, that water vapour is invisible. This means that children seek alternative explanations for condensation than its having come from the air. Occasionally I have had children tell me that condensation on the inside of the classroom windows is because 'the window is leaking'. Most children do see this as being implausible but they still find it difficult to accept that water vapour is condensing out of the air. The most common explanation I have heard them advance is that 'the water comes from our breath'. To some extent this is actually true as some of the water vapour in the room may well have been exhaled, but, when I have questioned the children, it usually becomes clear that their explanation is not the scientific one. Instead they usually believe that there is water in their breath and that somehow this comes directly through the air to the window. They almost never have any notion that in all but the very driest atmospheres, there will be some water present in the air and that it is in the form of a gas; water vapour.

Allied to this is the frequent belief that condensation is as the result of steam. It is easy to see where this idea comes from. Condensation is often associated with steamy environments such as kitchens and bathrooms. Many science books make matters worse by suggesting that teachers demonstrate condensation by boiling a kettle and holding a mirror, or similar, close to the spout. Even if the safety risks of this demonstration are addressed, it is still an inappropriate one as it reinforces the misconception just referred to. It is also conceptually dangerous as it tends to lead to the children believing that steam is the gaseous form of water, in fact, in my experience they are sadly often taught that this is the case. Steam is actually tiny droplets of liquid water suspended in the atmosphere, hence, unlike invisible water vapour, it can be seen and therefore, unsurprisingly, becomes the children's logical explanation for condensation. Even the 'breath theory' mentioned above is often linked to steam. Children have frequently told me that there 'is steam in our breath'. Again, this is not surprising as, when they exhale on cold days, they see water vapour in their breath condensing as mist (essentially cold steam which, just like clouds, fog etc. is actually droplets of liquid water suspended in the air).

Challenging the children's ideas

The particle model above can be used to help explain these phenomena but first it is useful to carry out some practical challenges to the children's misconceptions.

Firstly, it is important to address the notion that evaporated water has gone somewhere other than into the air. Letting the children evaporate water from a container that they have tested and agreed is impermeable is an important first step. Likewise this should be done in a manner that does not involve an extra heat source. (Later the children might investigate what affects the rate of evaporation but at first the fact that extra heat is unnecessary should be emphasized.) This activity will address the fact that the water has gone somewhere; the challenge is then to show where it has gone to.

This can neatly be tackled in combination with introducing condensation. The point here is to show condensation happening in a manner that can't be explained with reference to the steam theory, so, obviously no kettles! I prefer a demonstration using a container filled with ice cubes. Care must be taken with this. Traditionally this would have been done with a glass container but nowadays most schools sensibly will not let the children work with glassware. However I have found that the activity also works with hard plastic containers. It is sensible however for you to try out various safe containers first to ensure they work. It is a good idea if the container has been demonstrated to be impermeable but, to make extra sure that the children do not think that the condensation has come from the ice cubes, I make them from water with vegetable food colouring added to it. If you leave the container in the classroom, condensation will soon form on it but it might be best to set up the demonstration in a room without people in it to challenge the belief that the condensation comes directly from exhaled water. This will at least show water appearing on the exterior of the container that couldn't possibly have come from inside the container.

The next step is to combine evaporation and condensation using a self-contained system such as a plastic bowl with a little water in it and cling film stretched over the top. Again, it is useful if the children test the cling film and see that it is impermeable. Then, when they see water droplets condensing on its interior, the only place the water could have come from is within the bowl itself. If the children are sure that the droplets are not splashes from shaking the bowl, then they are one step closer to accepting a scientific model of evaporation and condensation.

This can then be discussed with reference to the particle model.

In any liquid, the particles are always moving about, a fact the children will be used to from the model introduced in Chapter 8. Evaporation occurs because some of the particles at the surface of the liquid manage to move about so much that they shake themselves loose and become a gas particle. (I prefer not to talk of energy at all as, in my experience the concepts involved are beyond all but the most very able of primary school children. I've never found this a problem as there are always alternative mechanical models.) Extra heat will speed this process up, as it makes the particles move faster, but it is not necessary.

In condensation, it is the reverse process. If the water vapour particles meet something colder, they slow down and begin to stick together, forming droplets of liquid water.

Once again, it helps if the children pretend to be particles and act the processes out.

Conceptual step three – applying a simple particle model to understand chemical reactions

Teaching primary school children about chemical reactions may seem very ambitious and indeed, if one were to plunge straight into such concepts, the children would very likely be confused, often because of the very problems already discussed in this chapter. However, if children have gone through the sort of conceptual progression already outlined, then I have found that they readily connect with the learning involved in understanding relatively simple chemical reactions.

The most important point about chemical reactions is that new substances are formed in them, either through particles combining to make new particles, or through particles separating to make new particles. All the changes in materials discussed so far have not resulted in new substances being made, only in the rearrangement of the existing substances.

Eliciting ideas

Eliciting the children's ideas is once again very easy in the sorts of contexts outlined at the beginning of the chapter.

For example, one material I frequently give the children to work with when they are acting as forensic scientists is health salts. The children may be conducting a variety of tests on the salts and it is easy to engineer that they have to add a substance to it that will result in a chemical reaction (water is the most straightforward). It is easy then to ask the children what they think is happening and to explain what they think the fizz is.

Equally, the prehistory context opens up excellent opportunities for discussing reactions. For example, when we were learning about possible ways that prehistoric peoples might have lit their fires, I asked the children what they thought was happening when materials burnt. Also, when we were making pottery, it was easy to ask the children what they thought had happened to the clay after it was fired.

It is important to remember that, by this stage, the children should be becoming adept at using their simple particle model to explain many of the everyday phenomena they encounter. This means that when eliciting their ideas, it is perfectly in order to ask them to express their ideas in terms of the particles they have learnt about.

Be warned!

Research warns of various misconceptions associated with understanding chemical reactions but the most serious ones have already been discussed and relate to the problems children have with thinking in terms of particles generally. As mentioned above, the sort of conceptual progress outlined in Chapters 6–8 will help challenge these misconceptions and give the children simple models to make sense of the phenomena they encounter.

None the less, the processes of chemical reactions are complex and their mechanisms are effectively invisible to the children, even if they can see the results. This often means that, even if they can suggest that particles are changing, they struggle to outline exactly how. I have found that in these cases some direct teaching is usually required but that, if the outlined conceptual progression has been followed, that this works very well, the children assimilating the new knowledge to the frameworks they have built up with little difficulty.

Challenging the children's ideas

So, in this case, we need be less worried about challenging the children's ideas as this will largely have been done previously. However, it is important to be careful with the form the direct teaching should take. Some common examples follow.

Adding water to health salts etc.

I have found that, when children have been introduced to materials in the manner outlined in the previous three chapters, they readily accept that the bubbles produced are a gas. This is a vindication of the approach, as children usually find gases difficult to understand. The question then arises as to where the gas has come from. Once again, if the children have connected with the models introduced they have no difficulty in realizing that the gas hasn't been spontaneously conjured into existence. Interestingly, I find children often suggest the gas has come from the water and that, when they are questioned it becomes apparent that many of them by this stage are aware of the fact that water is made from a combination of two gases. This is a simple factual error and the scientific reasoning behind it means that the children will readily accept that the gas has actually come from the health salts. In fact, I have often been delighted to see children do even better than this. Frequently I have found that children who had earlier been introduced to gases through the use of fizzy drinks machines, as mentioned in Chapter 7 suggested that the 'bubbles were fizz like carbon dioxide'. These children have often said things like 'there is no carbon dioxide in water normally, otherwise it would be fizzy, so the gas must have come out of the health salts'. This sort of reasoning once again shows just how well children can respond, if we are thoughtful about how we progressively introduce them to concepts. When your children are reasoning like this, it is easy to introduce them to what actually happens in the reaction. As usual, I find it helpful to the children if they act out what goes on, as in Figures 9.4 and 9.5.

Some children who are gas particles and some who are 'salt' particles, link themselves together and act as health salt particles clumped in granules. Other children stand by, acting as water: always in contact with each other but moving around one another. (Not to scale.)

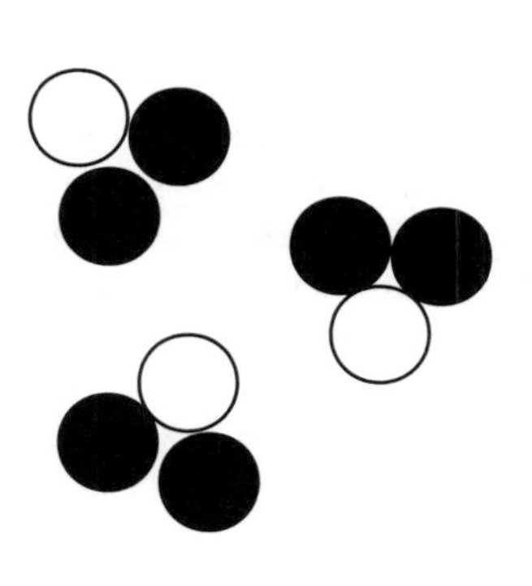

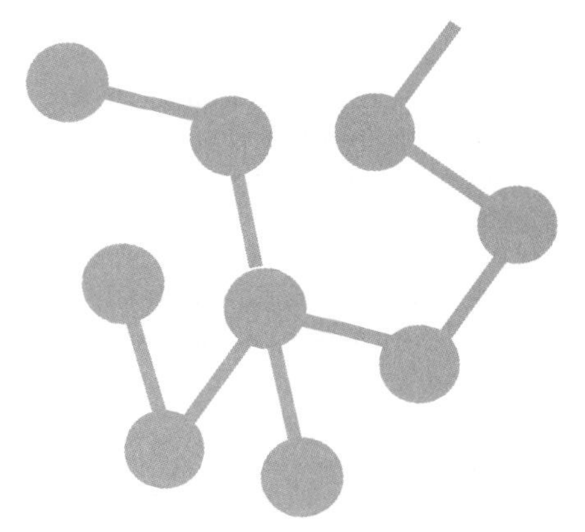

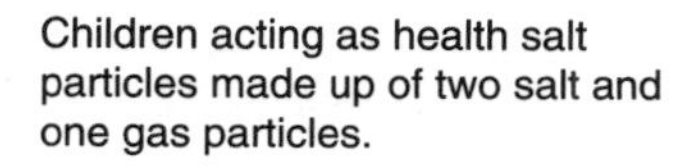

Children acting as health salt particles made up of two salt and one gas particles.

Children acting as water particles

Figure 9.4 Diagram of children acting out health salt and water particles prior to reacting

The water is poured onto the health salts. In the reaction, the gas particles are liberated into the atmosphere and the remaining 'salt' particles are dissolved into the water. (Not to scale.)

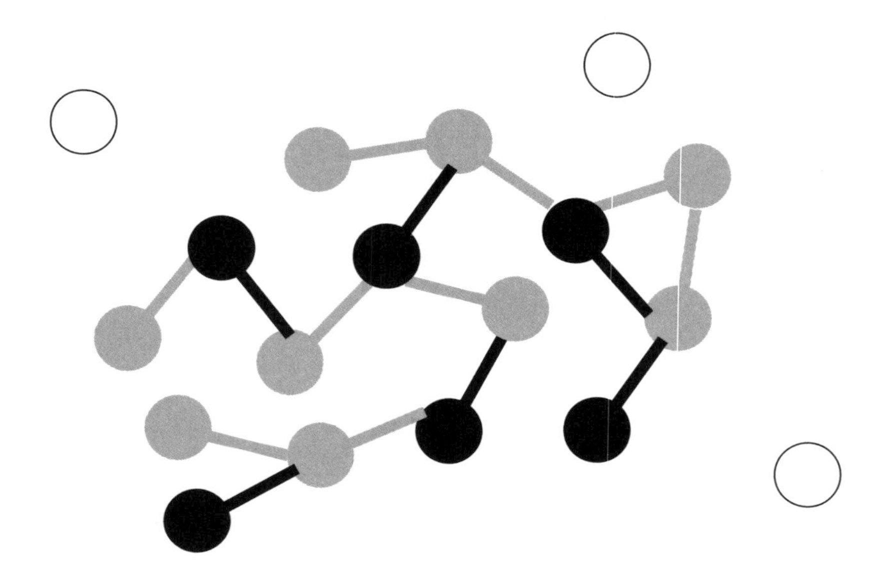

The salt particles have now dissolved and become a liquid as part of the salt solution. The gas has been liberated in the reaction and has gone into the atmosphere

Figure 9.5 Diagram of children acting out health salt particles reacting with water

There can still be problems however. This sort of activity is a great one for checking that the children understand that gases, as materials, are made of particles and consequently have mass. You may find that, if you ask the children is there a difference between the weight of the water and health salts before and after the reaction, some claim the weight is heavier afterwards. This illustrates the very common misconception amongst children that gases have a sort of 'anti-gravity' effect and make things lighter. Common related problems are children believing that inflated balloons full of air are lighter than deflated balloons; or that a 'flat' bottle of pop will be heavier than one that still has the fizz in it. Hopefully, if the children have got to grips with gases as materials, they won't have these misconceptions but it is worth checking.

Burning

Burning is a complex reaction, if one is to consider all the precise chemical changes involved but, at primary school level, the key point is that oxygen combines with particles in some form of fuel and that this releases heat and may produce various other products such as smoke, or even water. I find that the intricacies of this make it difficult to act out. I concentrate on helping the children appreciate the following key facts:

1. Some form of heat is required to start the burning reaction.
2. Some form of fuel is required, for example: wood on a bonfire; vaporized wax in a candle flame; or vaporized petrol in a car engine.
3. Oxygen is necessary as it is the oxygen combining with particles in the fuel that is the chemical reaction we call burning.

Take care because some children confuse melting (a purely physical change, as discussed in Chapter 8) with burning, which is a chemical reaction (where the particles involved are combined into new substances, rather than remaining the same type and being simply physically rearranged).

Firing clay

This is a straightforward reaction. The particles in clay are arranged in flat sheets. When clay is wet these sheets of particles can slide around over each other, allowing it to be easily manipulated. However, when the clay is fired, the particles in the different sheets are firmly joined together making it become rigid and more brittle.

Rusting

Like burning, oxygen is involved in this reaction, although in this case, water is also required. The oxygen combines with the iron to form a new substance; iron oxide, or, in everyday language, rust.

10

Developing a primary school model of thermal insulation

Introduction

As was mentioned in Chapter 8, the concepts introduced to children through their initial work on the topic of materials and their properties, particularly if a simple particle model of matter is used, will stand them in good stead when it comes to explaining many everyday phenomena. Thermal insulation is a good example of this. With a simple particle model, children can not only make observations regarding thermal insulation, but can also develop an understanding of why it occurs. In turn, the concepts developed will help the children when it comes to understanding other phenomena, such as light.

Where this topic fits in

The concepts covered in this section are all likely to be covered in key stage 2. However, they are not particularly complex and I have found they can be covered early in the key stage. There is an extra advantage in doing this as early coverage of these concepts will help the children when they come to tackle other, more complex topics in other areas of their science learning, such as light and electricity.

A capacity for thermal insulation is just one specific, if conceptually more difficult, property of materials. Accordingly it is important that, before dealing with this area you are thoroughly familiar with the principles outlined in Chapter 6.

Links to other topics

This topic builds heavily on the concepts of Chapter 8, which at key stage 2 are typically extended in three particular areas: thermal insulation; magnetic behaviour; and electrical conductivity. This section will concentrate on the thermal insulation aspects. Magnetism is addressed in Chapter 11. Electrical conductivity is best covered in the context of work on electricity and is dealt with in Chapters 16 and 17.

The initial concepts discussed in this section link closely with concepts involved in looking at changes in materials. When helping the children learn about solids, liquids and gases, you will find it useful to look at the relevant parts of Chapters 6 and 8.

A context for the topic

The concepts of this section can be tackled by a variety of imaginative topics.

One of my favourite and most popular topics with younger key stage 2 children is the creation of two sets of mythical creatures: the ice trolls and their friends the lava dragons. To an undiscerning eye, an ice troll looks merely like a small pop bottle, filled with frozen water and with a head stuck to the lid. A lava dragon appears similar though based around an aluminium drinks can filled with warm water (see Figure 10.1). Both creatures live in Iceland, the trolls amongst the frozen glaciers and the dragons around warm volcanic vents. They are friends but they have a problem: the trolls melt and get poorly if they try to visit the dragons; whilst the dragons chill down and get poorly if they try to visit the trolls. Some trolls have heard that if they wrap themselves up it will help them stay frozen but many other trolls think this is bad idea. The trolls want to know what the children think and ask for their assistance to help find out if wrapping up would work. The children can then investigate if the same tactic would help the lava dragons.

This is obviously a great context for learning about thermal insulation but it also allows the sort of general exploration of materials required for the first conceptual step in this section.

Figure 10.1 An ice troll, proudly wearing his new coat goes to meet his friend the lava dragon

For older key stage 2 children I have often used contexts that build on grouping and classifying materials and involve the children in testing the properties of materials for a variety of uses. As mentioned in Chapter 8, some of the most successful contexts have involved survival situations. For example, once, while studying the Spanish Armada as part of the children's work in history, I set up a shipwreck situation where the children imagined they had been wrecked on a deserted Scottish island and had to survive until help arrived. The children were given access to a range of salvaged and naturally available materials and had to decide how they would use them to survive. The children then conducted tests on the materials to see if their suggestions would have helped them stay alive. For example some children had suggested making mattresses to stay off the cold ground and carried out an investigation comparing cloth stuffed with bracken to cloth stuffed with dried seaweed, to see which was the better thermal insulator.

Conceptual step one – noticing and discussing: that some materials are good at keeping warm things from cooling down; and some everyday instances where materials are used to stop cold things from warming up

The various national curricula tend to suggest that thermal insulation should be dealt with at key stage 2 and this makes sense in that the older children will grasp the rather

Table 10.1 Conceptual steps and example learning objectives for developing a primary school model of thermal insulation (the first conceptual step refers to background that can be developed in key stage one; all other learning objectives are suitable for use at key stage 2)

Conceptual step	*Example learning objectives*
Noticing and discussing: • that some materials are good at keeping warm things from cooling down; • some everyday instances where materials are used to stop cold things from warming up.	Children will: notice that some materials are good at keeping warm things from cooling down; identify some everyday instances where materials are used to stop cold things from warming up; (N.B. At this level I do not set formal learning objectives regarding the explanation of how thermal insulation works. The conceptual explanations of thermal insulation are best left until the children are rather more experienced.)
Using thermal insulation to solve problems.	learn that materials can both slow: • the warming up of cooler things; and the cooling down of warmer things;
Learning about thermal insulation in action.	be able to identify everyday uses of thermal insulation;
Building a concept of what thermal insulation does.	learn that cold is a relative absence of heat; learn that heat always moves from hotter to less hot, never the other way around; understand that thermal insulation works by slowing the transfer of heat from hotter things to less hot things.

abstract principles more easily. However the younger children are bound to begin discussing how some materials 'are good at keeping things warm' and so it will be helpful to outline ideas for dealing with the topic at an appropriate key stage 1 level. I do not set formal learning objectives regarding explaining thermal insulation for key stage 1 children but instead encourage the children to begin to think about the topic by noticing everyday examples of where materials are used to help keep other materials warm, or to help stop other materials from heating up. I have found that carefully addressing and discussing these points at key stage 1 has helped reduce problems with misconceptions later on.

The aim here is merely to get the children beginning to think about thermal insulation. In fact, I have found that many able and slightly older key stage 1 children can cope with simple investigations regarding thermal insulation but, in a crowded and time-pressurized curriculum, I reckon that it is a poor use of time to extend them in this area and that the time is better devoted to other more suitable concepts, leaving the more advanced work until key stage 2.

It is worthwhile initiating discussion however, and this is best done as part of the sort of work on thinking about the properties of materials, introduced in Chapter 6.

Eliciting ideas

You can use the kinds of strategies that were introduced in Chapter 6 to elicit the children's ideas but it may be inappropriate to go too far in challenging the ideas, as this is likely to involve introducing concepts that will tax the children at this level. I try to focus the thinking on two main areas.

First I encourage the children to notice that some materials are good at keeping warm things warm. This observation tends to be readily advanced by even young children, especially in the sort of dressing up context considered in Chapter 6.

The second observation that I encourage is for the children to notice some everyday instances of materials being used to help stop cold things from warming up, for example cool boxes being used to slow the warming of picnic foods, or thermos flasks being used to keep cold drinks from warming up as quickly.

Be warned!

Many of your children are likely to be confused and believe that thermal insulation actually warms thing up. In effect they believe that insulation generates heat, although they are most unlikely to articulate this misconception in such words.

In addition, the concepts involved in insulation slowing the heating of cool objects are more difficult for the children than those involved where insulation slows the cooling of warm objects, despite the fact that, scientifically speaking, the process involved is identical. The children are likely to perceive some materials as intrinsically cooler, rather than appreciating that they are in fact poorer thermal insulators and may suggest that cold objects should be wrapped in 'cool materials' to help keep them cold. Equally, if they still believe that insulation actually generates heat, they will probably believe wrapping up cool objects to be a bad idea as they will see this as likely to heat them up. (You should note that further detail regarding these difficulties is given in the following sections.)

Challenging the children's ideas

Challenging such a misconception by systematic practical investigations may be beyond children at this stage but you can still do something to help their conceptual development. Notice the exact phrase I used above, *to notice that some materials are good at keeping warm things warm.* The key part of the phrase is *keeping warm things warm.* Insulation will not warm up an already cold object, but it will slow down the cooling of an already warm object. The key thing to encourage the children to observe is that these materials only help to keep warm things that are *already* warm. (Note that objects that actually generate heat, such as the children themselves, will actually get warmer, if thermally insulated. This does cause the children conceptual problems and how to address this is further discussed in the following sections.)

Likewise, regarding insulation slowing the warming of objects, it is also best to avoid challenging the children's beliefs in much practical detail, as the explanatory concepts are likely to prove too abstract for them. Instead I concentrate on simply looking at some everyday instances of thermal insulation being used to slow the warming of cool objects in the hope that this will sow conceptual seeds that can be followed up in key stage 2.

Conceptual step two – using thermal insulation to solve problems

Ideally your children will have covered the concepts outlined in step one as part of their work on materials in key stage 1. If this is the case, they may need a little revision of the ideas involved but it should be not be difficult to soon be setting them some practical problems involving thermal insulation.

Eliciting ideas

If you use learning contexts similar to those outlined above, it is easy to get the children thinking about and communicating their own ideas. For example, the ice trolls directly ask the children if they think wearing a coat will help the trolls not melt as quickly when they visit the lava dragons. Or, to give an example from another context, if space-wrecked on a desert planet, the children may have to come up with ideas for staying cool during the day and warm at night.

Hint!

When eliciting the children's ideas in practical contexts such as these, let the children think practically. When they are discussing their ideas, give them access to the sorts of materials they will be investigating, for example: a pile of materials salvaged from their wrecked ship; or the trolls might bring some substances that they think could be made into coats. The children will be able to reason more effectively if they have the stimulus of the actual materials and the process may also allow the identification of misconceptions, such as problems stemming from the fact that some materials seem to feel warmer than others.

Be warned!

You are likely to have to contend with many misconceptions in this topic. The fact that many children believe insulation actually generates heat has already been noted in the previous section, along with some basic strategies for challenging this misconception. Even if revisiting the topic in key stage 2 you are likely to find that some children still hold this view.

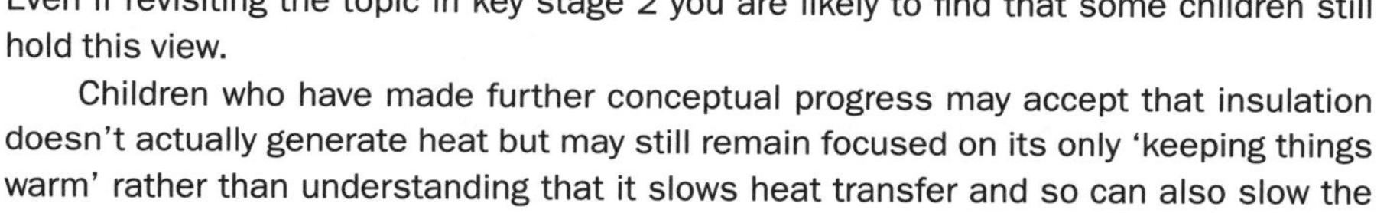

Children who have made further conceptual progress may accept that insulation doesn't actually generate heat but may still remain focused on its only 'keeping things warm' rather than understanding that it slows heat transfer and so can also slow the warming up of cooler objects.

Challenging the children's ideas

Setting problem-solving investigations for the use of thermal insulation will effectively challenge both these ideas. For example, testing to see if coats will slow the melting of ice trolls will show both that insulation does not generate heat and that it can slow the warming of cooler objects. The basic concepts will need to be further refined but this becomes much easier once these foundational problems have been tackled.

Hint!

I have found it preferable to start the children with an investigation that involves using insulation to slow the warming up of cooler objects rather than one involving using insulation to slow heat loss from warmer objects. This will challenge the common misconceptions listed above right from the start. Indeed, beginning with an investigation into insulation slowing heat loss may only reinforce these misconceptions.

Conceptual step three – learning about thermal insulation in action

This is now a relatively straightforward task and, in my experience, the children have little difficulty in identifying a range of uses for thermal insulation.

Eliciting ideas

The easiest way to do this is to ask the children to list the uses they can think about, find in their home etc.

Be warned!

Since this is now a straightforward exercise in the application of existing concepts there should hopefully not be too many difficulties. However, when discussing the differing materials used for insulation you may encounter the common misconception that some materials are always colder than others. For instance, children have often made statements to me such as, 'metal is colder than wood'.

Related to this problem is the potential confusion caused by the different ways in which thermal insulation can work. The problems are exemplified by a question one ten-year-old once asked me: 'If people get foil blankets to stop them getting cold after a marathon, why aren't mountaineering clothes made out of tough foil?'

Challenging the children's ideas

The notion that some materials are always colder than others comes from a lack of understanding of how heat can be transferred by conduction. A metal chair leg will feel colder to your touch than a fabric chair back as the metal will conduct the heat away more quickly from your hand. The easiest way to challenge this idea is to measure the temperature of the different materials. I usually do this with computer data logging probes. I am always a bit apprehensive that slight inaccuracies in the probes might lead to the materials appearing to be at different temperatures and so I try this out before letting the children make any measurements but so far I've never encountered any problems.

The question asked by the child about the foil clothes is a very intelligent one and the answer lies in the fact that heat can be transferred in three different ways: radiation, conduction and convection. Radiation is where heat is directly 'beamed' off from a heat source and can take place even in a vacuum (this is how the Sun's heat reaches us across space). Conduction is where heat is transferred directly through continuous or touching materials, such as from an electric hob to a metal saucepan. Convection is where gases or liquids move past a heat source and carry heat away with them, such as currents in the water in a kettle spreading heat right through the water (it's not just the water next to the element that heats up). Unfortunately, examining these concepts is rather outside key stage 2 science but, as the example question shows, it is as well to bear them in mind. I do not attempt to teach these concepts but I am ready to discuss examples with the children, should they raise the issue. For example, after a marathon, our muscles are very warm and radiate lots of heat; a foil blanket is good insulation in such circumstances as it is effective at reflecting radiating heat. A mountaineer, on the other hand, is likely to lose most heat through convection, in other words the wind blowing over his body and taking heat away with it. Foil clothes would not be effective insulators in such circumstances. The best insulation for mountaineers is several layers of fleecy fabric, topped by a wind-proof outer layer. The fleecy fabrics trap air which is warmed by body heat and, since the air is trapped, convection cannot take the heat away. Finally, an example of a material providing an effective thermal insulator against conduction might be the plastic grip on the handle of the mountaineer's metal ice axe. Grasping the metal would lead to heat loss through conduction but gripping the plastic greatly slows this down.

Conceptual step four – building a concept of what thermal insulation does

Once the children have carried out such problem-solving activities and examined everyday uses of thermal insulation, steps can be taken to help them extend their understanding of exactly how the insulation works. This is a rather trickier phase than the last conceptual step but is very important as the children will commonly hold misconceptions that will compromise their understanding of the property of thermal insulation. Indeed, even if they have successfully completed the problem-solving activity, they are still likely to be confused on several important points. Some of these points are rather more abstract and so, depending upon the abilities of your children, they might be quite challenging in a lower key stage 2 context and so be better left for a later look at materials.

Eliciting ideas

A good way to initiate discussion and find out the children's ideas is to compare what happens to a cup of cold juice from the fridge and a cup of warm tea when they are both placed in the classroom. I also usually prepare a second pair of cups but insulate them. The children are likely to have no difficulty in correctly predicting that the juice will warm up, whilst the tea will cool down. They will also by this stage hopefully have no difficulty in realizing that the insulation will slow this process down. I also find that they can correctly predict that the juice will not keep warming up until it boils and that the tea will not keep cooling until it freezes. It is then interesting to ask them what their explanations for these phenomena are.

Be warned!

The children's responses are likely to alert you to several common misconceptions. The children will usually appreciate that the tea cools because it loses heat and they will usually correctly state that the insulation 'helps keep the heat in'. I often find that they have more problems with the warming juice and here they often see the role of the insulation as 'to help keep the cold in'. This reflects the common misconception of thinking of 'cold' as a positive entity, rather than just the absence of heat. Similarly, in the case of a winter coat, for example, they will sometimes talk of the role of insulation being to, 'keep the cold out'.

Another common source of confusion is thinking about fridges and freezers, which are usually interpreted as working by generating cold, whereas they actually work by taking heat away from the materials placed in them.

Challenging the children's ideas

Unfortunately this is one of the rare instances in primary science where you will have to resort to telling children the answer, rather than their being able to work it out themselves; either through a practical test or a thought experiment. However, I have always found that when I have followed a cycle of activities similar to those outlined above, the children have not had difficulty accepting these final concepts. I concentrate on establishing the following four facts:

1 there is no such thing as *cold*, cold is only a relative absence of heat;
2 heat only ever travels from hotter to less hot materials;
3 thermal insulation slows this travelling from hotter to less hot;
4 and once everything is at the same temperature, there is no further heat transfer.

Thus the discussion activity of the cups of juice and tea can be explained as is outlined in Figures 10.2 and 10.3.

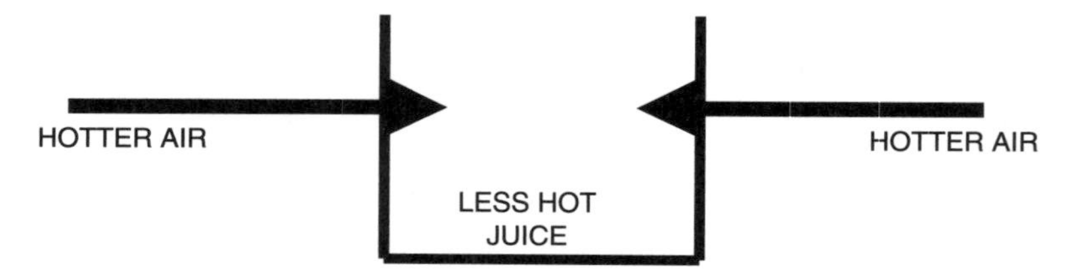

Here heat travels from the hotter air to the less hot juice

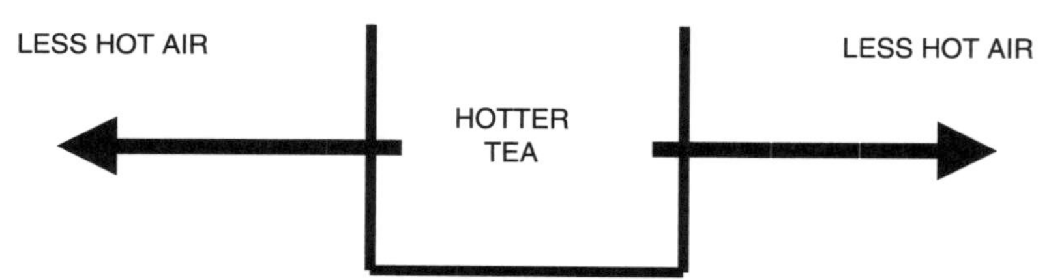

Here heat travels from the hotter tea to the less hot air.

In each case, the temperature of the liquid and the air will eventually become the same (room temperature) at which point the heat transfer will stop.

Figure 10.2 Heat transfer with uninsulated cups

Heat travels in the same direction as before but the insulation slows the rate of transfer

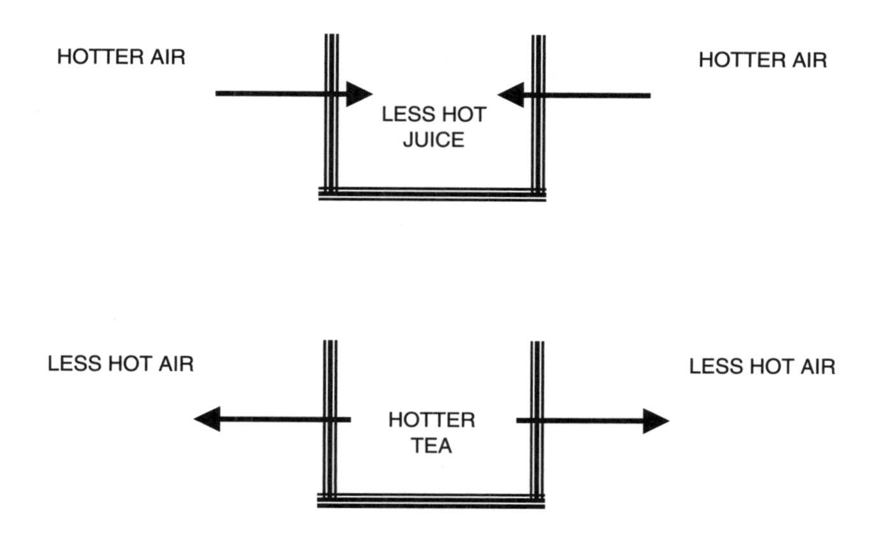

Once again, in each case, the temperature of the liquid and the air will eventually become the same (room temperature) at which point the heat transfer will stop. The only difference to the process as shown in Figure 10.2 is that the insulation slows the process down and so the time taken to even out at a new room temperature is longer.

Figure 10.3 Heat transfer with insulated cups

Hint!

Computer data logging equipment is extremely useful in contexts like these. For example, after the discussion of the elicitation exercise with the uninsulated and insulated cups of cold juice and warm tea, I usually get the children to set the exercise up again but with data logging equipment to measure and graph the temperatures of the four liquids. Asking the children to predict what the graph will look like will give you a useful insight into their thinking, as well as being general good practice. Later, the actual graphs will allow the children to compare the rates of temperature change and, crucially to see that, eventually, each cup reaches the same temperature and stays there.

You can now discuss with the children how fridges work by pumping chemicals through their frames which take heat away from the materials inside that is then transferred to the air by a radiator on the back of the appliance.

Once the children have grasped these concepts you can also do some useful reinforcement of learning, and have some fun, by getting the children to spot how everyday language can reinforce misconceptions about thermal insulation and heat transfer. For example, 'That's a nice warm jumper you're wearing', could become, 'That's an effective thermal insulator you're wearing', or 'Always shut the fridge door to keep the cold in', should become, 'Always shut the fridge door to keep the heat out'.

11

Developing a primary school model of magnetism

Introduction

Magnetism is a very popular primary school topic, which is no wonder given the children's fascination with the seemingly magical effects of magnets; after all, no less a figure than Einstein traced his fascination with science to having been given a compass to cheer him up during an illness. You will certainly not have difficulty motivating your children with this topic. It is also good news that, compared to many other science topics, magnetism does not typically have difficult, entrenched misconceptions. None the less, there are some pitfalls and sometimes it is easy for the topic to be not much more than a vehicle for the children to practise their science skills, without really focusing on magnetism itself.

Where this topic fits in

The first two conceptual steps outlined for this topic are easily accessible to younger children and should typically be covered at key stage 1. The later steps should be tackled at key stage 2 but, even then, they are not particularly complex and can usually be covered early in the key stage.

Links to other topics

Magnetism has important links to several other primary science topics.

It is closely linked to materials and their properties. This is not just in terms of the properties of materials (capable of being magnetized, attracted to magnets etc.) but the work on a particle model of matter discussed in Chapter 8 will also be helpful when the children come to develop a model of what magnetism actually is.

Magnetism is also often linked to the topic of forces. Here the link is not an explanatory one, but the fact that magnets can exert forces (pushes and pulls). Magnets are particularly useful in demonstrating forces acting at a distance.

Crucially, magnetism has links to the topic of electricity. In fact, electricity and magnetism are both manifestations of the electromagnetic force. More experienced primary school children may begin to explore these links by observing how electric currents affect compass needles and by building electromagnets. There will also be useful reinforcing links in the children's explanatory models between the two topics. For this reason I always ensure my children have studied a simple introductory topic on magnets, before they come to build up a model of how, for instance, a battery works.

A context for the topic

As I've already stated, children get very enthusiastic when learning about magnets and, from a purely motivational point of view, it would be appropriate to just let the topic stand as 'magnets'. Of course, as was discussed in Chapters 3 and 4, effective learning is not just about motivation and it is possible to help your children's learning by giving them a problem-solving context that lets them use their science skills to test their own ideas. A great idea for a magnets topic is to have the children design and make their own magnetic games. Even simple games such as magnetic fishing, magnetic treasure hunts, or magnetic race tracks will allow young children to explore and test the properties of magnets.

For the later conceptual steps with older children, I rarely have a separate magnetism topic but instead cover the relevant conceptual steps in other topic areas. For example, I have already described how I often use survival topics to help the children learn about materials and their properties (as discussed in Chapter 8). As part of such a topic, it is easy to introduce magnets: for example, the children may have to use materials to make a homemade compass. Such work is an excellent way of addressing the learning under conceptual step three. The aspects of conceptual step four can then be covered as part of the children's work on electricity. (The relevant aspects of electricity are discussed in Chapters 16 and 17.)

Table 11.1 Conceptual steps and example learning objectives for developing a primary school model of magnetism (the first two conceptual steps can be tackled in key stage one; the latter two steps are best left until key stage 2)

Conceptual step	*Example learning objectives*
	Children will:
Learning that magnets can attract some other materials and not others.	learn that magnets can only attract some metals; notice that magnetism can act at a distance;
Noticing that magnets can both attract and repel other magnets.	learn that magnets have two poles and that different poles attract, whilst like poles repel;
Building a model of how magnetism works.	understand a simple model of how some materials can be magnetized and demagnetized;
Noticing some simple links between magnetism and electricity.	learn that an electric current has a magnetic effect.

Conceptual step one – noticing that magnets can attract some other materials

This is a straightforward step and one that easily fits in with the example topic of designing magnetic games.

Eliciting ideas

This can be done as part of the problem solving in connection with designing games. For example, the children might have the problem of how to get their cardboard fish to stick to a magnet in a magnetic fishing game. It is easy then to ask them for their ideas as to what sort of materials might be used.

> ***Be warned!***
>
> It is likely that you will encounter some confusion here. Many young children, for example, will assume that any metal will stick to a magnet. However such confusion does not represent a major misconception and is easily challenged.

Challenging the children's ideas

This is easily done by letting the children test a variety of materials. The children will soon find that only certain metals are attracted to magnets. Here there is a potential problem depending on how advanced the children's knowledge of materials is. The metal most strongly attracted to magnets is iron. However, alloys containing iron, such as steel, will also be attracted to a magnet. Less well known, and perhaps less common, nickel and cobalt will be attracted to magnets. Such discrimination is likely to be beyond the young children who can otherwise perfectly get to grips with the conceptual step. With younger children, the key point will be that only some metals are attracted and it may be the case that you then simply have to tell them that iron is most strongly attracted to magnets; but that metals containing enough iron will be also; as well as a few other metals (nickel and cobalt).

Interestingly, although iron is the easiest metal to magnetize, it also loses its magnetism easily. Steel is harder to magnetize but retains it's magnetism much better. This is why your school magnets will be made of steel.

> ***Be warned!***
>
> There is further scope for confusion here. I remember one bright six-year-old predicting that a two pence piece would not be attracted to a magnet because it was copper, not iron. This child obviously had some useful scientific background but was shocked when she discovered that the coin did stick to the magnet! Of course, I didn't initially offer an explanation but asked the girl what she thought. 'Perhaps there is iron in it,' she

suggested. It was a good introduction to talking about alloys. These days, a pure copper coin would not only be likely to be worth more than its nominal denomination but, as copper is relatively soft, would also wear out relatively quickly. It is not a surprise then to find that 'copper' coins are not pure copper but made from alloys that may, or may not, be magnetic. The UK two pence coin the girl was investigating was actually made of steel, even if coloured, and referred to as 'copper'.

Be careful, too, as I have seen children become confused when, for example, plastic coatings or varnishes stop magnets sticking to what they consider are metals. Also, even if there is some iron in a metal alloy, there may not be enough to make it magnetic, as is the case with some stainless steels. Once again though, such confusions are easily sorted when the child is encouraged to observe a little more carefully.

Conceptual step two – noticing that magnets can both attract and repel other magnets

Once again, a theme of magnetic games will soon introduce the children to these observations.

Eliciting ideas

To get discussion going, I sometimes show the children a set of 'floating magnets' and ask them what they think is going on. (Floating magnets are easily available from educational suppliers. They consist of several, pierced, disc-shaped magnets that can be threaded onto a vertical dowel rod. If the magnets are threaded onto the dowel with like poles facing each other they will repel each other and appear to magically float in mid air.) If the children have already played with magnets, they usually quickly come up with the idea that 'the magnets are pushing each other apart', or something similar.

Be warned!

The problem here is not that young children won't readily appreciate that the magnets can sometimes attract each other and sometimes repel each other; but in what their explanation for this will be. In fact, the scientific explanation for exactly how magnetism works is extremely complex and involves quantum physics that is certainly beyond the primary school classroom. Even so, it is possible to develop a simplified primary school model. This is outlined under conceptual step three but it draws on a particle model of materials and is unlikely to be suitable for many younger children.

Challenging the children's ideas

So, if an explanatory model is likely to prove too complex for this age group of children, it is best to concentrate on the observations of what the magnets are doing. Through these the children will hopefully appreciate that:

1 Magnets have two different 'ends'. (They will readily adopt the technical terminology of *poles*.)
2 Different poles attract each other.
3 Like poles repel each other.

An explanatory model can then be left until the children have a little more experience of materials and their properties.

Conceptual step three – building a model of how magnetism works

The example of a survival context discussed earlier, where the children have to attempt to make a home-made compass makes an ideal introduction for discussing magnetism in more depth with older children.

Eliciting ideas

If the children are set the problem of how they can find a given direction such as north, they will come up with a range of suggestions and a compass will certainly be one of these. They may have less idea about how to actually make one but, if you give them access to magnets, you will often get imaginative suggestions. It is possible to simply hang a bar magnet on a thread but it is much more effective to make a floating needle compass. (To do this, the children have to magnetize a needle by stroking it repeatedly, in the same direction, with the same end of a strong magnet. Once magnetized, the needle can be floated on water, resting on a thin bed of expanded polystyrene.) You may well have to give guidance on this but children enjoy the task and are fascinated by the results.

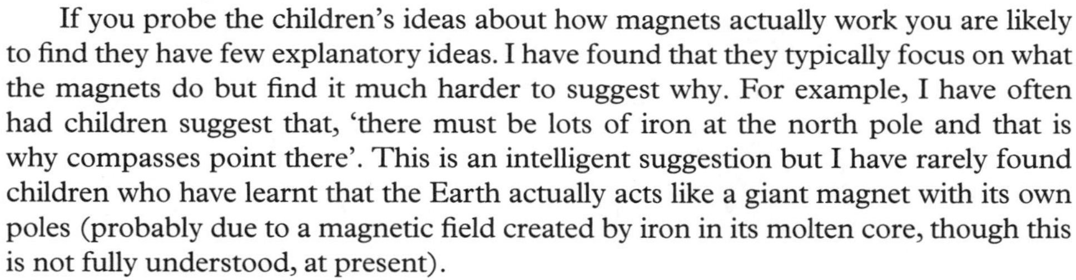

If you probe the children's ideas about how magnets actually work you are likely to find they have few explanatory ideas. I have found that they typically focus on what the magnets do but find it much harder to suggest why. For example, I have often had children suggest that, 'there must be lots of iron at the north pole and that is why compasses point there'. This is an intelligent suggestion but I have rarely found children who have learnt that the Earth actually acts like a giant magnet with its own poles (probably due to a magnetic field created by iron in its molten core, though this is not fully understood, at present).

If you think about it, you will realize that something odd is going on here! Like poles repel; so, how come the north pole of a magnet is attracted to the north pole? The answer is that what we usually call in shorthand the 'north pole' of a magnet, should really be termed the 'north-seeking pole'. In other words: relative to the Earth's magnetic north pole, the north-seeking pole of a magnet is actually a 'south pole' and is therefore attracted to the Earth's magnetic north pole! You may decide such distinctions are beyond your learners but I remember one sharp eleven-year-old who quizzed me about this. At any rate, it is very unlikely that the children will come up with such suggestions on their own! Even a child who comes up with the sensible, if incorrect suggestion, about lots of iron at the north pole will most likely struggle to suggest how it attracts a compass needle. The best suggestions I have encountered have been from children who have been introduced to a simple particle model of matter, as discussed in Chapter 8. Often such children have advanced suggestions about it being something to do with the particles in magnets and this makes the best foundation for a primary school model of magnetism.

Challenging the children's ideas

If the children have been introduced to a particle model of matter, as discussed in Chapter 8, they will be familiar with the fact that all materials are made up of particles. From this beginning, I encourage the children to think of the particles exerting different pushes and pulls on one another. These pushes and pulls are like the way in which the two poles of magnets can attract, or repel each other. This is certainly the case and without the electromagnetic force, everyday materials would literally fall apart. The bizarre quantum things that particles get up to are beyond primary school children but I have found that they will accept the fact that the way particles behave in materials means that they act like lots of little 'mini-magnets' inside the material. Technically these are called magnetic domains. However, because the mini-magnets are all jumbled up, the pushes and pulls cancel out and the material doesn't act like a magnet, as in Figure 11.1.

In some substances, however, most especially iron, it is possible for the magnetic domains to be lined up in a particular way so that the pushes and pulls reinforce each other, as in Figure 11.2. If this happens it has been magnetized. This can be done either by stroking with another magnet, or by using electric current to line up the magnetic domains.

This is a very basic model and, in my experience, most of the children have exaggerated, simplistic views of what the particles in the magnet are doing. Even very

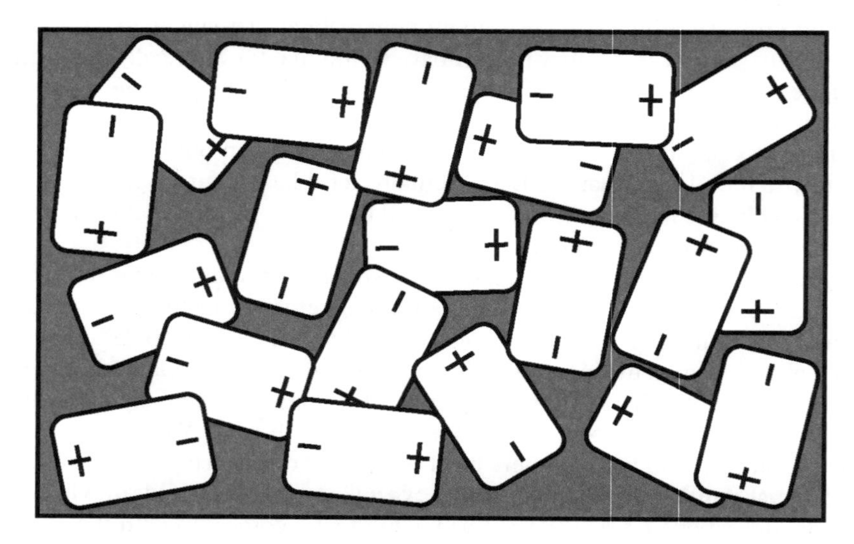

Figure 11.1 Iron with all the magnetic domains jumbled up so the pushes and pulls cancel each other out

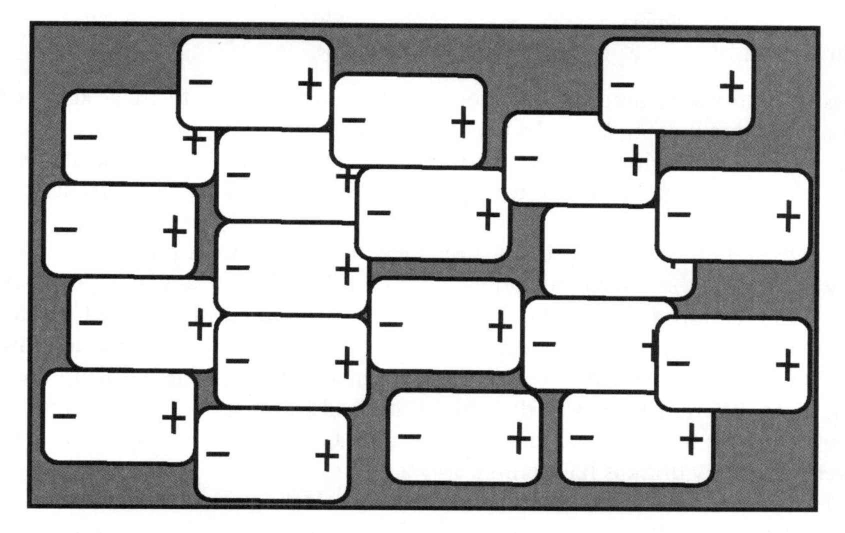

Figure 11.2 Iron that has been magnetized so that the magnetic domains are arranged so the pushes and pulls reinforce each other

able children present inaccurate views of what is actually happening; one boy suggested to me that 'the electrons must all be at one side and that makes it repel or attract'. But the model does give the children a working explanation for what goes on in a magnet and if you can encourage your children to think as scientifically as this, you should be very pleased. You can also use the model to help reinforce good practice in looking after magnets as, not only can the magnetic domains be lined up, they can also be jumbled up once again, which will de-magnetize the magnet. Common ways that this might happen in the classroom are:

- dropping the magnet on a hard surface;
- heating the magnet up, such as by leaving it on a radiator;
- jumbling magnets up together (good primary school magnet sets will have storage boxes that prevent this).

Conceptual step four – noticing some simple links between magnetism and electricity

Some national curricula specify that these concepts be left until after primary school but they are often included in key stage 2 materials and are certainly within the grasp of most key stage 2 children.

I have found it best to address this conceptual step as part of the children's work on electricity, typically after the concepts addressed in Chapter 17. If you do this, the learning can form part of a technology topic designed to allow the children to apply, consolidate (and you to assess) their knowledge of electricity.

Eliciting ideas

This is an interesting area in that most children are unlikely to make any connection between magnetism and electricity from their own experience (even if much of the modern technology they encounter on a daily basis depends on it). Accordingly, a simple activity is required to get the children thinking and discussing their ideas.

I ask the children to construct a simple electrical circuit. (A bulb is not necessary but it will increase the resistance of the circuit and slow the rate at which your batteries go flat.) The children should be given a button compass and be asked to see what happens when the compass is brought near the wires. You can encourage them to explore the effects of moving the compass closer to the wires; further from the wires; placing it at different points of the circuit; and adding batteries. (Remember to be careful to choose bulbs to match the 'push' (voltage) of the batteries.) Then ask the children what they think is happening.

The children will certainly notice the current affecting the needle but they may be rather baffled as to why. This illustrates the importance of taking care of the progression of the children's thinking; of thinking carefully about the order of your topics; and of striving to provide the children with explanatory models appropriate to their understanding.

I have found that children tend to be fairly baffled unless they have connected with the concepts and models outlined earlier in this chapter; with the sort of particle models of matter in Chapter 8; and with the models of electricity discussed in Chapters 16 and 17. I have found such children sometimes striving to come up with ideas but just not having the experience to let them get far; typically they say things like 'the power of the electricity pulls the compass'. Sometimes they come up with better founded, if still incorrect, suggestions, such as the child who said to me, 'it's the metal in the wires that pulls the magnet thing' (referring to the compass needle).

However, I have found that, if children have had the sort of foundational experiences referred to above, they can often manage to advance more sophisticated ideas. Children have said things to me such as, 'the electricity must be changing the particles in the wire to make it like a magnet', or, 'the electrons in the wire are minus and the compass must be positive, so it pulls the needle'. Whilst these simple explanations may not reflect the true complexity of what is happening, I regard them as excellent thinking for children at this stage and a suitable foundation to build upon, as the children progress their scientific experience.

Challenging the children's ideas

Once again, the full scientific explanation is beyond the children but I think it good progress if the children are beginning to think as outlined above. I concentrate on helping them to think as follows:

- electric current (a flow of electrons) has a magnetic effect;
- and that there is a parallel between how the poles of a magnet attract and repel and how positively and negatively charged particles attract and repel each other.

The children's experience can also be extended by their making simple electromagnets by wrapping coils of wire around, for example, a four-inch nail. (Iron would be even better but steel nails are likely to be more easily available.) Instructions for this sort of activity are found in many science books but you should note the safety instructions given as, to get the electromagnets to work best, you may need to use batteries of stronger 'push', for example, 6V 'lantern' batteries and there may be a risk of connections becoming quite hot. There is scope for excellent investigative and technological work here, however, and the activity also allows the children to see electric current temporarily rearranging the magnetic domains in steel so that it becomes magnetized.

Be warned!

When I was a child it was normal practice to let children experiment with magnets, paper and iron filings to see the patterns of force around a magnet; its magnetic field. Whilst this is certainly an intriguing activity, there are two problems. Firstly, explaining the fields will be beyond the children, though they can certainly make some basic progress using the concepts mentioned. Secondly, and much more serious, is the fact that the iron filings are not only messy and a curse in how they coat the magnets but they are a safety risk, particularly if they get into the children's eyes. If you are going to let your children examine such phenomena, make sure you use filings in sealed containers that are specially designed and approved for such use.

12

Developing an initial understanding of light

Introduction

Light is often seen as a straightforward and fun primary science topic with many useful links to other curricular areas from examining the role of light in various religious festivals to building shadow puppets. However it is actually rather more complex than it appears and children hold many misconceptions about what light is, how it travels and its connection with sight.

Where this topic fits in

The learning involved in this section is relatively accessible and can easily be covered in one unit at any stage of key stage 1.

Links to other topics

It is useful to have done some basic work on materials and their properties before covering this topic as this will be helpful to the children in their explanation of some of the phenomena they explore, for instance how the properties of some materials make them appear to be light sources because they are shiny and reflect a lot of light.

A context for the topic

There are lots of children's stories that involve light and dark and that can be used as a stimulus for learning in this topic. However it is both easy and enjoyable to make up your own stories and scenarios with the children and this can also make for some good cross-curricular links. A context that I have found a great favourite with the children is the *Thunderbirds* TV series. Despite the series' age and the strings showing on the puppets, I have had great fun, and great learning successes, with young children acting out the adventures of *International Rescue* in a variety of contexts. Light has been no exception and the scenario of having to rescue some foolish older children who had been trapped in a collapsed tunnel provided an excellent stimulus for eliciting and challenging the children's ideas about light and dark. The scenario was also used for a variety of cross-curricular links from imaginative writing to exploring issues of safe and responsible behaviour and was also used to reinforce learning the children had encountered in their last science topic, which had been materials and their properties.

Conceptual step one – identifying light sources

When considering this area, it is important to define what is regarded as a light source. In some subject knowledge books a distinction is made between primary light sources, such as the Sun, that actually produce light and secondary light sources, such as the Moon, that give off lots of light but do not actually produce it, only reflect light from other light sources. I have found this distinction to be very confusing to primary children, especially the younger ones for whom the concept of reflection is difficult anyway. Instead I concentrate on light sources as only being objects that actually produce light.

Eliciting ideas

There are many ways in which the children's ideas on light sources might be elicited. In the context of the above scenario, the children were asked to imagine what problems they would have trying to stay alive in the tunnel until *International Rescue* arrived to save them. The children quickly appreciated that it would be dark in the collapsed mine and this lead on to discussing questions such as: Can you see in the dark? Why was it dark in the tunnel? and What might the children have that would give light? To back up the exercise and provide further stimulus for discussion, I had examples of the backpacks the children had with them when they were trapped. These contained various things that might be used as light sources such as torches, a mobile phone, matches and a camping stove; as well as examples of objects that might

Table 12.1 Conceptual steps and example learning objectives for developing an initial understanding of light (all of these conceptual steps are best covered in key stage 1)

Conceptual step	*Example learning objectives*
Identifying light sources.	Children will learn: to identify commonly encountered light sources; how to conduct a test to ascertain if an object is a light source, or not; why items may appear to be light sources, even when they are not;
Understanding that darkness is the absence of light.	that darkness is the absence of light;
Formulating a simple model of how light relates to vision.	that light is required for vision; that creatures that can sense in complete darkness use senses other than sight.

be confused with light sources, such as shiny metal plates and brightly coloured waterproofs. The packs and the objects were discussed with groups of children allowing further insights into their ideas.

Be warned!

Children often show misconceptions regarding light sources. They readily identify light sources such as light bulbs, candles, and fires and are usually quick to spot that objects such as mobile phones, TVs and computer monitor screens can also give off light. However, many children become confused by shiny, reflective objects such as mirrors or reflective road safety armbands. Bright and light coloured objects can also sometimes be regarded as light sources. The discussion of why it was dark in the tunnel raised suggestions such as, 'there was no Sun so it was dark', and it is true that children usually have no difficulty in accepting the Sun as a light source. I have often found though, that young children suggest the sky and sometimes even clouds as light sources and I have even heard them advance quite sophisticated (if misplaced) reasons for this, for example one child told me, 'clouds make light because it's light even when it's cloudy'. Stars can often highlight the children's incomplete understanding; sometimes children view them as light sources simply because 'they're shiny', but I have often had children make statements similar to 'stars don't make light because it's dark when they are out'. Hardly surprisingly, the Moon causes great confusion and, in my experience, the youngest children simply cannot grasp the concept that it is not producing light but merely reflecting the light of the (often unseen) Sun.

Challenging the children's ideas

The key to challenging the children's misconceptions is for them to appreciate that the test for an object being a light source is if it can be seen in the dark. The rescue context is very helpful in this respect as the scenario helps the children link light sources with vision and begin to appreciate that darkness is the absence of light (which will be discussed more fully, following). When I have asked the children how we could test whether the various objects would make light for the trapped children, they have often immediately suggested taking them into the dark. If a dark area is constructed, the children can then bring the various objects into it and, if they can be seen, they are light sources and must actually be making light. If the objects can't be seen, they must not be making light and must only reflect light. You should be aware though that the concept of what actually happens in reflection is challenging for young children and that, for many children at key stage 1 level, their understanding is likely to be something on the level of 'the mirror doesn't make light, it is only shiny if there is light shining on it'.

Such tactics work extremely well in challenging misconceptions such as believing mirrors to be light sources but they won't work for the Moon, sky, clouds or stars. For the youngest children the concept of the Moon reflecting sunlight may simply be inaccessible and you may find that it is best for them at this stage to simply accept that the Moon is a light source. If you are working with older or more capable key stage 1 children you may find that they can grasp this fact and ideas for challenging their ideas can be found in Chapter 14.

Challenging the notion that clouds are light sources can be tricky. Appealing to the fact that there can be clouds at night may be confusing as it is not completely dark at night, a feature the children may actually attribute to the sky or clouds! Instead I have found it better to concentrate on helping them understand that it is not completely dark at night, the proof of this being that they can see things, even if not very well. (Once again this relates to concepts introduced in the following conceptual steps.) Once the children accept that they can see at night, it is then possible for them to try and consider what the light sources may be at night: stars, lights of distant towns etc. and, at this point, exploratory play experimenting with what light sources look like from various distances can help the children appreciate that the stars are actually huge, bright Suns that are very far away. Regarding the clouds, I have found it important to give the children the opportunity to develop a model for how the Sun can shine through clouds, making it light even on a cloudy day. This can be easily done by allowing them to experiment with shining lights through various substances showing how some light can get through.

Conceptual step two – understanding that darkness is the absence of light

This concept is far from obvious to children but is linked closely to the other two conceptual steps described in this section. The example context provides a useful illustration of how to link the learning of each.

Eliciting ideas

The initial discussion about the trapped children would also help the children's ideas regarding darkness, particularly by posing the question, why is it dark in the tunnel? It is also useful to ask the children if they think that it would be possible to see in the collapsed tunnel. You will also gain further insight into the children's ideas if you extend the discussion to ask where else the children think it is dark and if they think that it is ever possible to see in the dark.

Be warned!

A context such as this is helpful to the children in forming ideas and they will often explain the fact that it is dark in the tunnel by saying that there are no lights or that the tunnel has collapsed and so the light can't get in. They are also likely to suggest that that the fact that there are no lights is why the children can't see. Such responses are encouraging but may give a misleading picture of the children's understanding and that is why it is important to ask the children where else they think it is dark and if they think that it is ever possible to see in the dark. These extra questions often uncover misconceptions. Typical responses include examples such as, 'it's dark at night', or, 'it's dark under my bed'. Many children will also claim they can see in the dark and may link this to factors such as eating carrots! Many will also suggest that there are creatures such as owls and cats that can see in the dark. All of these indicate an incomplete understanding of the concept that darkness is the absence of light and are linked closely with the fact that the children may never have, or only very rarely have, experienced complete darkness.

Challenging the children's ideas

Challenging the children's ideas in these respects is easy in parallel with devising a test for light sources and involves the creation of a dark area. The key point is that it must be completely dark i.e. let no light in at all. In the *Thunderbirds* context, I set the children the challenge of designing an area where *International Rescue* can train to work in the dark and where Brains, the organization's scientist, can test various light sources. By asking the children to create somewhere that is dark, we are effectively forcing them to look at how light can be kept out. It is important to be rigorous about this and constantly ask questions such as: Is there any light? Can you see any light? And how can we keep that light out? It is fun for the children to come up with their own dark area strategies but it is useful to note that blackout curtain material is readily available from fabric retailers and can easily be made into dark tents that, draped carefully over tables, will exclude all light and allow the children to experience complete darkness. Such opportunities can be extended to include simple investigations, for example, in my *International Rescue* scenario, Brains regularly asks the children to test a range of torches using a computer data logger to see which is the brightest. Once complete darkness has been experienced in this way, it is much easier for the children to understand that darkness is the absence of light and to appreciate that, if it is completely dark, they can't see.

My students have often worried about the children being frightened in such situations but I have never found this a problem. We as teachers may be managing the situation but we should be doing so in such a way that the children have ownership of the activities and I have found that when they are organizing matters themselves, far from finding the experience of the dark tent frightening, they actually take great delight in it.

This will certainly help challenge misconceptions such as it being dark under the bed, or at night. The children will appreciate that there is less light under the bed or at night but that it is not completely dark and it is important to link this to the fact that they can see; in contrast, if the dark tent is properly constructed, there will be no light entering it at all and the children will be able to see nothing when inside. However, the other potential misconceptions regarding 'seeing in the dark' must still be addressed. These cannot be directly challenged in a practical way but, after the activities above, the children will be more likely to modify their ideas through discussion of alternatives. Thus carrots can be discussed as containing vitamins and minerals that help keep eyes healthy but that won't allow miraculous powers! The idea that some creatures may be able to see in the dark must also be challenged and this makes a good bridge to the next conceptual step.

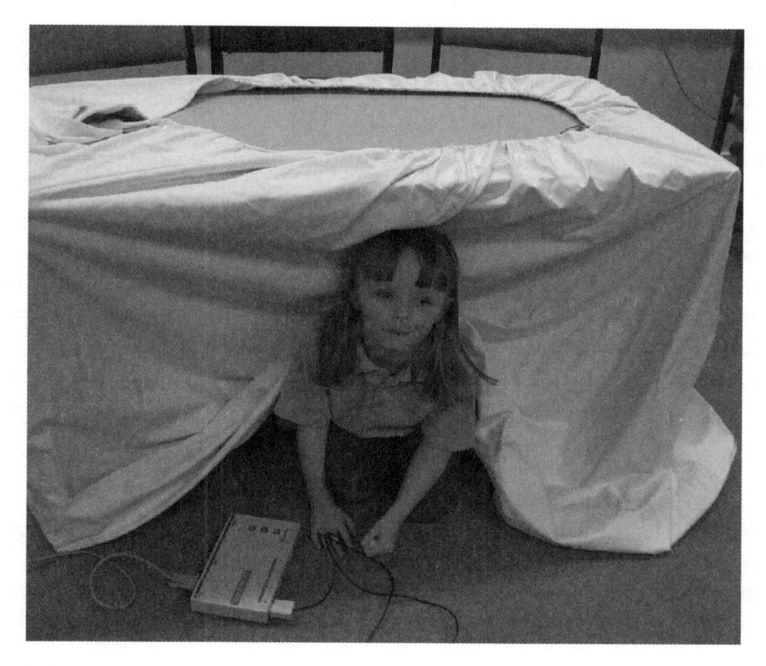

Figure 12.1 A young investigator emerging from a dark tent where she has been testing which light source is the brightest with a computer data logging system

Conceptual step three – formulating a simple model of how light relates to vision

How we see is often not considered as part of the key stage 1 curriculum and, indeed, complex models of how the eye works are not appropriate but, as the preceding discussion shows, a consideration of vision is unavoidable, if the children are to meet the learning objectives entailed in the first two conceptual steps; after all the 'light detectors' they will usually be relying on are their own eyes! Accordingly it is important that the children get the chance to reflect upon what they have discovered in this respect and to come up with a simple model of how they see.

Eliciting ideas

The preceding activities should already have done this to some degree through discussion of what is required to see in the tunnel but it is worth extending the discussion to ask the children how they think their eyes work.

Be warned!

This is a complex area and I certainly do not expect highly developed ideas from the children. There are a couple of points I find useful to check however. The first is to ensure that the children have now firmly accepted the idea that light is needed to see. Hopefully their own experiences in the preceding activities will have convinced them that they themselves need light to see but, as has been mentioned they may still believe other creatures such as cats, bats or owls can see in the dark. The other area I have found it useful to explore is to check if the children believe that seeing is an active rather than a passive process. Many young children believe that their eyes actually produce some kind of 'seeing rays'. They may believe this even if they accept that light is necessary to see; I remember a child once telling me, 'I need the light to power up my eyes so that they can look out and see things'. Such children find it difficult to accept that seeing is a passive process with light literally just going into their eyes through a hole in the middle.

Challenging the children's ideas

Dealing with the misconception that some animals can see in the dark is quite straightforward. The children can be encouraged to find out how some animals actually just have eyes that are very sensitive to low levels of light, such as cats. They can also explore how some animals may also use senses other than sight, for instance: owls have exceptional hearing; bats use echolocation; and some rattlesnakes sense the heat of their prey.

The active vision problem is more complex and a whole model may well be beyond the capability of some key stage 1 children but, in my experience, the misconception is often bound up with the children's common belief that they can see in the dark and so it is as well to be aware of it and to take what action one can. The crucial concept at this stage is that the children accept that they need light to see and thus they cannot see in complete darkness since darkness is the absence of light. Many children will then accept that it is the light going into their eyes that lets them see rather than 'seeing rays' going out and discussion of the pupil as a hole to let light in and how closing our eyelids mean that we cannot see as it stops light getting in may help reinforce this. This is easier when thinking about light sources as the children find it much easier to accept that they can 'see' the light from the torch and so believe that the light from the torch goes into their eyes. They find it much harder to accept that light is being reflected from non-light sources and also going into their eyes. Playing in the dark tent helps as the children will typically easily reach the conclusion that they can see non-light sources when they are 'lit up' by the light from light sources. Using shiny objects may then help begin to get across the fact that this is because light from the light sources bounces off the non light sources and goes into the children's eyes. You will find that some children at this stage have difficulty accepting this and in such cases it is certainly not worth pushing matters. However since the ideas can easily be explored through structured play activities, I always introduce the concept. It helps challenge the misconception of being able to see in the dark and introduces concepts that are to be crucial later on. Indeed I have often found that children who have been introduced to such ideas in key stage 1, even if they have not fully accepted them then, are quicker to grapple with the concepts in key stage 2 than children who have never encountered the ideas.

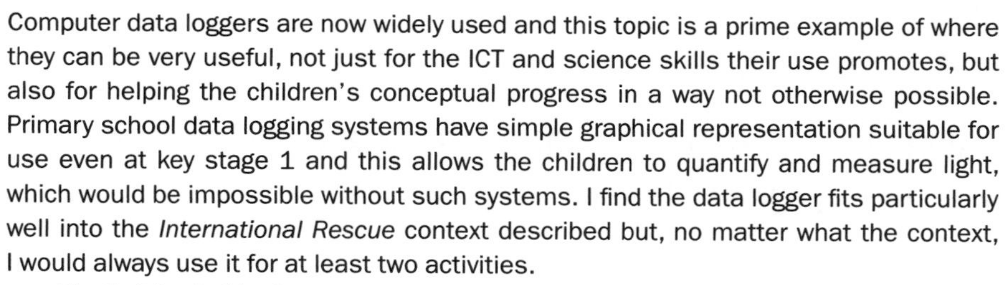

Hint!

Computer data loggers are now widely used and this topic is a prime example of where they can be very useful, not just for the ICT and science skills their use promotes, but also for helping the children's conceptual progress in a way not otherwise possible. Primary school data logging systems have simple graphical representation suitable for use even at key stage 1 and this allows the children to quantify and measure light, which would be impossible without such systems. I find the data logger fits particularly well into the *International Rescue* context described but, no matter what the context, I would always use it for at least two activities.

Firstly it is vital in demonstrating that darkness is the absence of light and can be used to show the children in a quantifiable fashion that there really is no light inside the dark tent. This also then reinforces the link between vision and light; the data logger shows there is no light inside the tent and the children cannot see there because of this. It is also interesting to allow the children the chance to play and experiment with using the data logger to measure different light levels in different places and at different times thus reinforcing that darker circumstances are associated with less light.

Secondly the data logger allows the children to carry out simple investigations. Testing various light sources to see which is the brightest has already been mentioned. It is also possible to test different materials to see how much light they let through and this is once again useful in reinforcing the concept that making things darker involves reducing the amount of light.

13

Developing a more advanced model for understanding light

Introduction

Even at key stage 2, light is frequently seen as a relatively simple primary science topic, typically consisting of some practical activities on shadows and reflection followed by discussion of how our eyes work. Such activities can be easily managed and can be fun for the children but they often leave common misconceptions unchallenged and do little to help children understand what light actually is. As in many topics, the key is to help the children build up an appropriate model for what light is that they can then use to explain the common everyday phenomena they encounter.

Where this topic fits in

The learning involved in this section covers quite a spread. Some aspects are quite basic and can be covered in lower key stage 2 classes, but some children may struggle with the more complex applications of the model. Accordingly, I have sometimes found it best to plan to visit the topic twice in key stage 2. This can be accommodated as, to some degree, the first three conceptual steps of this chapter are interrelated. Often, for example, it is possible to do some work on shadows before step one, as shadows are not particularly difficult and introduce experiences that may actually help the children connect with the model outlined in step one. This is an especially good strategy for schools that introduce light in three stages: once at key stage 1; once in lower key stage 2; and once in upper key stage 2. In such cases a unit on shadows in lower key stage 2 makes a good bridge between the key stage 1 work and the harder concepts that follow.

Links to other topics

As is often the case, if the children have a good grounding in materials and their properties, it will be helpful in their explanation of the phenomena they explore, for instance how the properties of materials affect how they reflect light.

It is important that children have had a thorough grounding in the concepts of how light travels and how shadows are formed before they study the concepts outlined in Chapter 14, *Earth and beyond*.

A context for the topic

Frequently this topic is taught as a stand alone topic with a title such as 'light and shadows'. This can work but it is much better if a more interesting context can be found. These can often have a strong cross-curricular element, perhaps even using an entirely different subject as a springboard for the children's work in science. For example, one of the most successful light topics I ever ran was based on a history topic on the theme of the Victorians. The children were studying Victorian engineering and it happened that there was a very fine Victorian lighthouse close to the school. This allowed a wider examination of lighthouses and how they worked, with various scientific investigations leading to some excellent design and technology work.

Conceptual step one – developing a basic model of what light is and how it travels

Chapter 12 introduced the foundation concepts for what light is: that it is produced from light sources; and that darkness is the absence of light. It is now appropriate to have the children extend these concepts into a basic model for what light is and how it travels.

Eliciting ideas

In the context discussed above, it was easy to elicit the children's ideas about light by setting them a practical problem to design a model lighthouse that would be as effective as possible. The very open-ended brief of 'as effective as possible' allowed for a great deal of open-ended discussion that gave considerable insight into the children's ideas. Examples of responses involved a wide variety of suggestions, some scientifically correct, and some scientifically incorrect, such as: using bigger bulbs to be brighter; using bigger batteries to make the light go further; making sure the material around the bulb was 'really transparent'; putting 'shiny stuff behind the bulb to bounce the light out'; and turning the classroom lights off so the 'lighthouse worked better'. This sort of discussion is a great way to find out what the children think. Listening to the responses alone is not enough; you also need to follow up with questions to clarify exactly what the children mean but, as the examples show, the activity allows you to identify children who have a good understanding of the concepts and those who have misconceptions. Best of all, such a task raises ideas that the children can then test for themselves.

However useful this sort of context is, you must also be prepared to ask the children what they think light actually is. Often they will happily talk of 'light' and engage in discussions such as the above without actually articulating what they believe light to be and it is important to know their ideas before providing them with a simple model.

Table 13.1 Conceptual steps and example learning objectives for developing a more advanced understanding of light (all of these conceptual steps are best covered in key stage 2)

Conceptual step	*Example learning objectives*
	Children will learn:
Developing a basic model of what light is and how it travels.	that light is comprised of tiny particles produced by light sources; that light particles travel away from light sources at incredibly high speed until they are blocked;
Understanding shadows.	that materials block light from passing through them to varying degrees; that shadows are caused when light is blocked by such materials; that the shapes of shadows vary depending on: the shape of the light-blocking object; the distance of the object from the light source; and the angle of the light source relative to the object;
Understanding reflection.	that, depending on the material that it meets, a light particle can either travel through the material; be reflected by it; or be absorbed by it; that the brighter a non-light source is, the more light it is reflecting; that some very smooth materials reflect light in such a way as to form an image;
Refining the model of how light travels.	that light spreads out as it travels away from light sources; that light travels until it is absorbed or reflected; to apply the model of light to be able to explain more complex phenomena such as why lights appear brighter at night and why not all lights can be seen from the same distance;
Understanding how we see.	that we see because light travels into our eyes, either directly from a light source; or indirectly, having been reflected off an object, and that this forms images in our eye.

Some of the examples given will be further discussed under succeeding conceptual steps but at first it is important to help the children formulate a basic model of what light is and how it travels.

Be warned!

Children often have little notion of how light travels. Typically they speak of light 'beaming down' from the Sun, light bulb etc., or they even speak of 'light just being there'. This is hardly surprising, given the incredible speed that light travels at, so fast that our eyes can never detect it moving; we don't, for example, see the light slowing spreading out from a bulb when we switch it on.

These problems also relate to the difficulty children have of articulating what light actually is. I have found some, typically younger, children equate light with light sources but this is unusual. Usually they do accept that something is coming from the sources but are unsure exactly what it is. Typical responses I have had are: 'it's bright stuff that comes out of the light sources' or 'it's bright, sparky stuff a bit like electricity'. I have often encountered children who have heard of light waves but they too are vague as to what this means in practice, typically making statements such as, 'the light is waving stuff made by the Sun and lights and fires and things like that'. This shows just how children struggle with the concept unless they are given a simple model to use.

This difficulty often confuses children about how light actually travels, for example, many believe that light travels only a certain distance and then 'runs out'. Also, it is often thought that light travels further at night than it does in the daytime.

Challenging the children's ideas

Some of the ideas mentioned above, such as those regarding transparency and reflection, are best addressed under the following conceptual steps. At this point, the important thing is to help the children develop a simple model of what light is. This is another example of where some direct teaching is beneficial, as the children will struggle to come up with a scientific model on their own. As always in good practice, this should still be based on the children's own ideas and observations built up through the sorts of activities discussed in Chapter 12.

Bearing in mind the typical children's responses, as discussed above, there are several key points that I stress in the model.

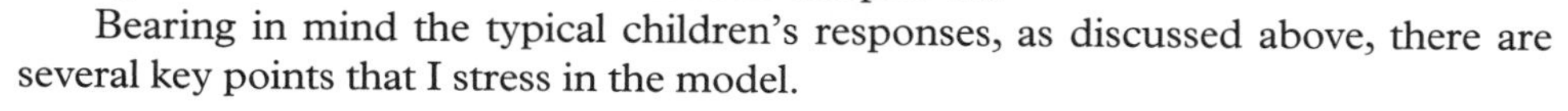

1 Light is produced in light sources. This builds on concepts already addressed in Chapter 12 but now the question of what light actually is must be addressed. Physicists consider light to have a dual nature; acting as both particle and wave. However, for primary school children (and many adults!) talk of light waves is hopelessly confusing and so it is the particle nature that I introduce to children and this is perfectly adequate to explain all the phenomena it is necessary to address at primary level.

2 And so the second point is that light consists of particles which, in everyday terms, are unimaginably numerous and unimaginably tiny. Curious children may well ask how these light particles are produced. If you encourage them to think of common features of light sources, they will frequently realize that very often light sources are hot and it is true that very hot things do often glow and produce light. I have also often been pleased that children who have been introduced to the concepts discussed in Chapter 9 frequently suggest to me that light might be produced by chemical reactions, which is also true. The actual quantum process which involves electrons jumping about in odd ways is much too complex for primary level but, if children make the sort of suggestions listed above, I am very pleased and content to leave it at that.

3. The more light particles a light source produces, the brighter it is.
4. Light particles travel away from light sources in all directions (though, as will be discussed following, they may be blocked) and they travel very fast; very fast indeed, at roughly 3 hundred thousand kilometres per second! Encourage the children to think what evidence there is for how fast light travels. The example of us not seeing light spread out slowly from a lamp when we switch it on might be helpful here.
5. The light particles travel in straight lines. Again it is important to encourage the children to see if they can spot this themselves. Questions such as why can we not see round corners might be helpful here. The work on shadows in the next conceptual point will also be helpful.
6. The light particles travel until something stops them. This means that some of the light our eyes detect has come from such great distances as to be difficult to imagine. Encourage the children to think about how far light might have travelled. The link they should have made between light and seeing (as discussed in the last chapter) should help them here. If we can see something, light from it must be reaching our eyes; what are the furthest things we can see? How far away are the stars and galaxies we can see in the sky?

This simple model will help the children when they come to explain some of the common phenomena they encounter that are associated with light, for example, shadows.

Conceptual step two – understanding shadows

As discussed above, it is sometimes desirable to switch this conceptual step with step one.

Eliciting ideas

Shadows are such a common phenomenon that it is not difficult to draw them to the children's attention and ask them what they think causes them. In the lighthouse topic, for example, we visited the lighthouse on a sunny day and it was easy to point out the shadow of the tower and ask the children what they thought caused it. We also noticed that the shadow had changed direction and size during our visit, which provided a useful stimulus. Back at the classroom, I asked the children to sketch out their ideas as to how they thought the shadow might change during the day, which gave me a valuable insight into their ideas about how shadows are formed.

Related to this, we also discussed the different materials the lighthouse was built of and why. This led to a good opportunity to hear the children's ideas about the opacity of materials and how this affects how much light that passes through them.

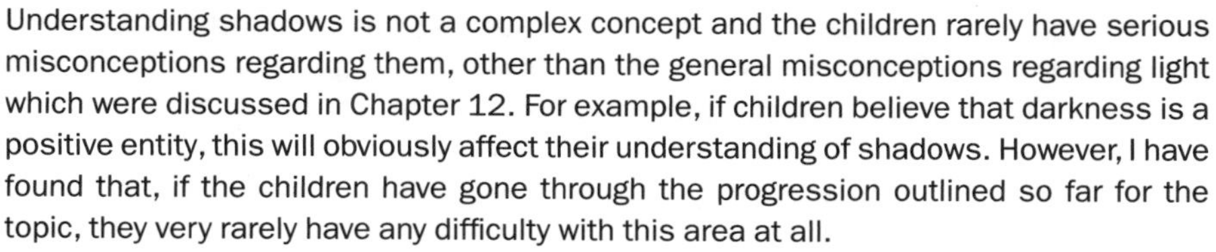

Be warned!

Understanding shadows is not a complex concept and the children rarely have serious misconceptions regarding them, other than the general misconceptions regarding light which were discussed in Chapter 12. For example, if children believe that darkness is a positive entity, this will obviously affect their understanding of shadows. However, I have found that, if the children have gone through the progression outlined so far for the topic, they very rarely have any difficulty with this area at all.

One minor, but common, misconception that I have encountered is that children often believe that it is only opaque objects that cast shadows. They don't appreciate that even an object normally considered transparent, such as a window, can still block some light and cast a faint shadow.

Also watch out for problems with terminology. I have often found that younger children confuse 'shadow' and 'reflection' using the words interchangeably. This does not reflect a deep misconception but can cause confusion. I usually treat shadow formation and reflection separately, which helps avoid any such difficulties.

Challenging the children's ideas

The activity on how the shadow changed provided an excellent lead into investigative work where the children tested out their ideas on how the shadow changed with model lighthouses in the playground. This led in turn to some groups carrying out some quite sophisticated investigations to identify the relationship between the position of the light source and the size of the shadow

Regarding opacity, the children also made predictions about how much light would get through different materials and then tested this with a computer data logger.

Through these investigations, it was easy for the children to establish that shadows are formed when something blocks light from passing through it, and that the more light that was blocked the darker the shadow. They also learnt how shadow direction and size varied and appreciated that the formation of shadows must mean that the light particles travelled away from the light source in straight lines.

Conceptual step three – understanding reflection

Reflection is more complex than shadow formation but, once again, I have found that, if the previous conceptual steps are followed, the children connect with the concepts well.

Eliciting ideas

This can be as simple as asking children why some materials are shinier or brighter than others. Alternatively, if you are setting the learning in a wider context, this might provide a better, practical lead in to the children discussing their ideas. For example, when discussing how to make a lighthouse as effective as possible, some children suggested putting shiny materials behind the bulb to make it brighter. One child, for example, said, 'my torch has silver stuff behind the bulb that reflects the light out and makes it brighter'. This allowed the children to debate this idea and I asked them to make predictions as to whether this was the case, or not, and if they thought there would be a difference between various materials.

Be warned!

By this stage hopefully the more serious misconceptions relevant to reflection will have been challenged through the activities already mentioned.

Possible confusions in terminology have already been discussed and so the children should be clear that shadow and reflection are two separate things. This confusion does hint at another problem though, that of the children believing that a material only reflects light if an image is formed.

Sometimes I have found children curious or confused about how various materials are different colours and this requires careful consideration. Sometimes they just don't understand how things can be different colours but sometimes their reasoning is quite sophisticated, for instance, when they believe that dark, matt objects don't reflect light, as they are not bright and shiny.

Challenging the children's ideas

The advantages of a practical context are once more shown here. When the children debated their ideas as to whether placing some materials behind a bulb would make it brighter, they were perfectly set up for science investigations that once more made good use of the data logging equipment. This allowed them to appreciate that different materials did indeed reflect light better. Interestingly, they also discovered that even backing materials that were not bright or shiny reflected more light back than the bulb standing on its own. This reinforced the concept that not just bright, shiny materials reflect light but in this area it is also always important to make the link back to vision: if you can see an object and it's not a light source, then it must be reflecting light.

The problem of believing reflection only occurs when an image is formed is a relatively minor problem but I used to encounter it commonly. However, I have found that, once a programme is instituted that allows the children to test light sources in a dark tent, or similar, then they better appreciate that shiny objects are not light sources and so must just be reflecting light, whether or not an image is formed in them. It is also important that the children get the opportunity to explore what properties of a material do seem to be linked with an image forming in it. They often suggest shininess, which is true, as opposed to the material being matt. However, where they equate shininess with brightness, they may be surprised, as even dark materials can form an image; here the key property that they usually soon notice is smoothness.

At this level, it is not necessary for children to draw ray diagrams illustrating how reflections form.

Dealing with why materials are different colours is tricky. It is easy, even at primary level, to show children white light being split into the visible spectrum by a prism and they will be familiar with these as being the colours of the rainbow. Conventionally the different colours of light are identified as being of different wavelengths but this is too complex for primary children. However, even if the full picture of wavelengths, energy etc. cannot be addressed, I believe that it is not inappropriate to talk of the different colours being due to differences in the light particles, an approach that I have found children have coped well with.

The problem is that this approach introduces the process of colour addition. This is how our brain adds together different combinations of light colours to make others: for example red, green and blue light will add together to give white. This concept is quite difficult for the children to grasp, especially as the process of mixing pigments (colour subtraction) with which the children will be familiar with from art, is totally different. These concepts can be introduced to primary level children but, for the learning to be effective, it takes a great deal of time and some quite specialized equipment (such as high grade colour filters) and so I do not usually undertake this. Where children are curious and raise these ideas themselves, I normally restrict myself to:

- showing white light being split with a prism;
- explaining that different colours are the result of differences between light particles;
- explaining that when different colours of light enter our eye, our brain adds them together to make other colours (for example, the seven colours of the visible spectrum add together to make white light);
- therefore the reason that materials are different colours is because they reflect different combinations of light particles.

Conceptual step four – refining the model of how light travels

At this stage, it is appropriate to revise the model of light introduced in conceptual step one and add to it the concept of light particles being absorbed and the effects of their spreading out as they travel. In addition, this is a useful time to deal with the common misconception that light travels further, or, as some children express it, ‘works better’, or ‘gets brighter’ in the dark.

Eliciting ideas

The children’s ideas regarding these concepts can be elicited in various ways.

Evidence for the fact that light particles can be absorbed as well as reflected should have come up during the work on reflection. For example, in the lighthouse topic, the children tested different reflectors and noticed that some reflected light better than others. This had been done as part of a fair test, where each reflector had had the same amount of light shone on it. It was easy then to ask the children why they thought less light was reflected from some materials than others.

The lighthouse topic also provided an excellent way of exploring the children's ideas regarding the other concepts.

I asked them why they thought the lamps in lighthouses were so big and when, as was very often the case, a child replied it was, 'so that it could be seen better', or 'seen from further away', I asked them what their explanation for this was.

When initially discussing what would make a lighthouse most effective, the children themselves had raised the point that the lamp would be easier to see in the dark and so it was straightforward to ask them if they agreed with this statement and what reasons they had for their belief.

Be warned!

In my experience, children take quite readily to the concept of light being absorbed but the other two concepts can be trickier.

I used to find that many children suggested that the light from bigger and more powerful lights would shine further; when they said a light was more powerful, they envisaged literally it pushing the light further. Sometimes they even equated this with the batteries in torches, reasoning that bigger batteries, for example, 'powered up the light more so it goes further'. However, I have found that after I began using the conceptual steps outlined and helping the children with the model of light described in conceptual step one, this problem was less common, as the children were more likely to accept that light particles travelled until something stopped them. Even so, I still find children who will suggest that the reason we can see stars at such incredible distances is because 'their light is so powerful and goes for ages'. Take care here because children can make a scientifically correct statement such as, 'we can see stars so far away because they are really bright', yet actually mean by this the incorrect explanation that their light is somehow 'stronger' and so goes further.

It is easy to see why children believe that light travels further at night as, without lots of competing light sources, lamps, candles etc. do appear brighter. Sometimes their level of reasoning is most impressive. For example one child once explained to me that the reason light went further was because, 'it is like sound. Sound goes better in solids than gases and the dark is more solid than the light'. This child had excellent reasoning skills and a good knowledge of sound but had the misconception of thinking of darkness as a positive entity. If you have helped the children realize that darkness is the absence of light, as discussed in Chapter 12, then you are less likely to have this problem but everyday appearances are still likely to confuse many children.

Challenging the children's ideas

So far as absorption is concerned, the best way to challenge and develop the children's ideas is a reflection test such as the one described. By using the data logging equipment, the children could measure that some materials reflected less light than others. Then, since they accepted that the materials were opaque, they easily appreciated that the differences between the reflective properties of the materials were because they absorbed varying proportions of the light particles hitting them. From this it was easy for the children to appreciate that bright materials absorbed less light particles and reflected more light particles; whilst dull materials absorbed more light particles and reflected less light particles.

I have found that even children who understand that light particles will travel until they are stopped, are then at a loss to explain why, as their experience will correctly tell them, they can see a lighthouse lamp from further away than they can see a torch, hence it is no wonder that some fall back on the seemingly sensible, though incorrect, idea that some light goes further than other light. This is another occasion when direct teaching is likely to be necessary as the children are unlikely to work out the correct explanation as to why brighter light sources can be seen from further away, which is because the light particles spread out more the further they go from the light source. However, if they have a good grasp of the model of light introduced in conceptual step one and they are shown a diagram like the one in Figure 13.1, children will sometimes grasp the concept without any further prompting.

Another factor that the children should appreciate after their work on materials is that when the light particles are going through the air they are going through a material, not through empty space. This means that, even though the light particles are so incredibly small, some of them will hit gas particles, dust particles etc. and be absorbed, or bounced back, which will also contribute to the fact that the concentration of light particles will diminish as they move away from the source.

Sometimes children have mentioned how certain torches 'focus the light more', and they are quite correct in that some torches have lenses that do just that. In such cases the light particles are kept in a tighter stream and so their concentration is kept higher. Interestingly, on the lighthouse project, the children also noticed that the

Light particles spread out more and more the further they go from light source. This means that there will be a greater concentration of light particles at position A than at position B. Eventually the concentration of light particles will be so weak that the eye cannot detect them. This is why brighter light sources can be seen further away; the concentration of light particles that they produce is greater at all distances.

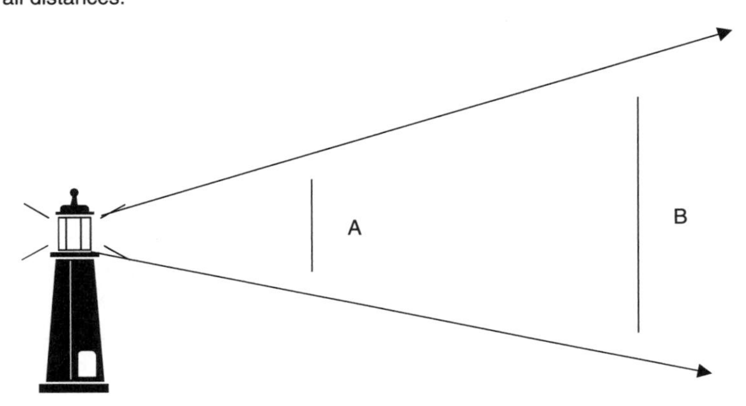

Figure 13.1 Diagram showing how light particles spread out more and more the further they go from light source

lighthouse lamps had lenses, as well as reflectors to concentrate the light more. If the children notice such features, there is no harm in discussing them but, just as with reflection, there is no need for the sorts of ray diagrams used in secondary schools.

Conceptual step five – understanding how we see

If your children have already gone through the conceptual steps outlined in Chapter 12 they should already appreciate the link between light and vision and have had some of the most common misconceptions regarding this challenged. However, vision is sometimes not emphasized when helping young children learn about light and, if this has been the case, you may find that problems emerge at this stage.

Eliciting ideas

The best way to do this is to ask the children to explain how they think they see various objects. Asking them to draw diagrams may be helpful but it is important to ensure you also ask the children to explain their diagrams as their interpretation of them may be quite different from yours. It is also important to ask the children to explain how they think they see both light sources and non-light sources. I also ask them to comment on some difficult objects such as a glass window and a black piece of card.

Be warned!

Research has uncovered many common misconceptions regarding vision. The notion that sight is an active process with 'seeing rays' coming from their eyes has been mentioned in Chapter 12 as has the mistaken belief that it is possible to see if it is (completely) dark. Even when the necessity for light is accepted, children can often articulate little more than, 'you need light to see with'. Sometimes, even with this understanding, children still see vision as an active process believing something along the lines of, 'we need the light to power our eyes'.

The good news is that I have found that, when children follow the conceptual steps outlined earlier and in Chapter 12, many of these misconceptions are challenged and laid to rest at an earlier stage. There can still be problems: the reason for including a window and black card in the pre-assessment list given above is because I have sometimes found children who believe that transparent objects let all the light particles through and reflect none, whilst they also sometimes think that black objects absorb all light particles and reflect none. On the whole however, the task is usually a relatively straightforward one of helping children consolidate a model of how we see.

Challenging the children's ideas

This model can easily be summarized in a diagram such as in Figure 13.2.

We see objects because our eyes passively detect light that enters them through the pupil. We see light sources as the light particles from them directly enter our eyes. We see non-light sources because light particles (originally emitted by light sources) are reflected off other objects and then enter our eyes. (N.B. at primary school level the subtleties involved with the angle of incidence equalling the angle of reflection are an unnecessary complication.)

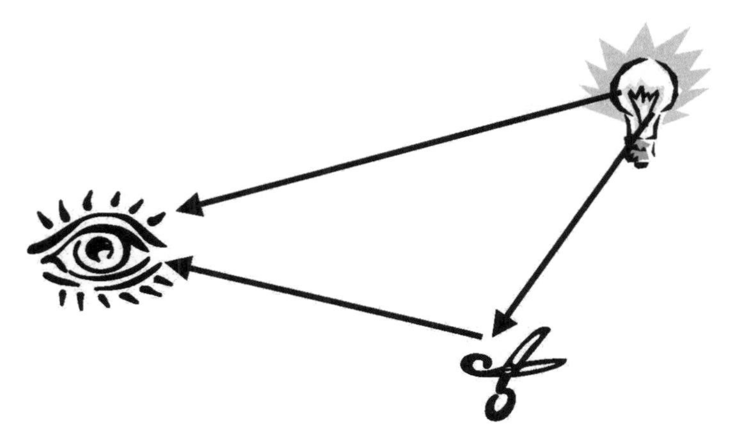

Figure 13.2 Diagram showing how we see objects because our eyes passively detect light that enters them through the pupil

I have often encountered primary school children being taught about the different parts of the eye and how to draw diagrams of it. This is not necessary. Although this sort of simply factual approach is not beyond the children, it is better to concentrate on building up the understanding as exemplified by the models discussed. Sadly, I have often noticed that this sort of factual approach often seems to be used as a substitute for proper, child-centred, investigative learning.

Hint!

Just as at key stage 1, computer data loggers are an invaluable aid in helping key stage 2 children learn about light. By allowing a quantifiable measurement of light they permit the children to carry out investigations into how light is reflected/absorbed by different materials and this is extremely useful in helping the children build up the sorts of models discussed.

14

Developing primary school models of Earth and beyond

Introduction

Children are fascinated with the universe beyond the Earth: the Sun, the Moon, other planets, comets, meteorites and the stars all fascinate them. Often this fascination is an imaginative one; connecting with ancient myths, or more modern speculation on UFOs and aliens. Often however, it is more scientific; a genuine desire to understand the phenomena they encounter. Both these strands make this an exciting and very rewarding topic to help children learn about.

Where this topic fits in

This is essentially a key stage 2 topic, the concepts involved being quite difficult for younger children. However there are some general foundations that can be profitably laid in key stage 1. Conceptual step one refers to this key stage 1 background; the other conceptual steps refer to work best undertaken in key stage 2.

Links to other topics

There are important links to other science topics and to other curricular areas.

A good understanding of light is essential if the children are to make sense of the concepts involved in conceptual steps two, three and four and I would never undertake such work unless the children had built up the models outlined in Chapters 12 and 13.

There are also many important cross-curricular links here. This chapter focuses on three major concepts traditionally included in science curricula: why we have day and night; why we have the seasons; and why we see the phases of the Moon. However these concepts are often included in broader topic work that examines the different objects of the solar system and perhaps discusses the stars, constellations etc. This sort of work is important and interesting for the children but it is not conceptually difficult and is often included in other subject areas such as geography and so is not discussed here.

A context for the topic

At key stage 1, the sort of background concepts discussed do not need a specific science context but can be addressed quite naturally through any number of cross-curricular contexts.

At key stage 2, one of the most successful contexts that I have used revolved around the theme of science fiction. As a means of stimulating the children's reading and writing, I would often have a genre theme that we would explore through reading, writing and visual media: school stories, horror, adventure, and detective fiction were all popular, and so was science fiction. One of the projects the children most enjoyed was designing their own fictional planets but, like much of the best science fiction (as the name implies), this had to be done along scientific lines. The children had to make lots of decisions about their planet involving such factors as its composition, size, day length, temperature, seasonality, climate; and then design the sort of eco-systems and species that lived on it. To make these sorts of decisions, the children had to understand the science behind the phenomena and studying their own planet was the perfect lead into this.

Table 14.1 Conceptual steps and example learning objectives for developing primary school models of the Earth and beyond (conceptual step one refers to background work undertaken in key stage 1; all of the other conceptual steps are best covered in key stage 2)

Conceptual step	*Example learning objectives*
	Children will learn:
Foundation observations at key stage 1.	that the sun rises in the morning and sets in the evening, marking the transitions from day to night; that there are seasonal variations and that these are not the same in all parts of the Earth; that the Moon is sometimes visible, both by night and day, and that it seems to change shape;
Understanding day and night.	that the Earth spins on its axis once every 24 hours; that when part of the Earth faces the Sun, it is daytime there; and that when it faces away from the Sun, it is night time;
Understanding seasons.	that seasonal change is typified by long days and short nights in the summer and the reverse in the winter; that these changes are more marked the further north, or south from the equator one goes; that the pattern in the northern hemisphere is the opposite to that in the southern hemisphere; that these changes are due to the Earth being tilted on its axis;
Understanding the phases of the moon.	that the moon appears to change shape on a regular cycle, which takes approximately 29.5 days; that this is due to the Moon being roughly spherical and thus certain parts of it always being in shadow.

Conceptual step one – foundation observations at key stage 1

These points can be covered through all sorts of work from simply looking out of the window at the sky and discussing what is seen, to using stories as a stimulus for discussion.

The concepts may seem very obvious but they are the first steps necessary for the children to understand the conceptual steps that follow and, if approached correctly, they can make a big difference to the children's later progress.

In this case the concepts are simply background observations and each will be discussed in turn without a requirement for the normal breakdown into eliciting and then challenging ideas.

Day and night, sunrise and sunset

This links closely to the children's work on light, as discussed in Chapter 12, as they should appreciate that the Sun is a light source. I concentrate on helping the children observe that:

1. The Sun rises in the morning and sets in the evening. At this stage undoubtedly many of the children will believe that it is the Sun that is moving, not the Earth but this can be challenged later.
2. It is daytime when the Sun has risen and is giving us light and night-time when the Sun has set. It is important for the children to realize that, even when the Sun is obscured by clouds, it is still giving us light. (Ideas regarding this are discussed in Chapter 12.)

The seasons

Here the children should be familiar with a variety of seasonal phenomena. These might include: flowering times of plants; some trees shedding their leaves; animals migrating, or hibernating; changes in typical weather; and festivals associated with the time of year. However, from a science point of view, I concentrate on helping the children observe that:

1. In the summer, days are long and nights are short and that in the winter, the reverse is true.
2. That when it is winter in the northern hemisphere, it is summer in the southern hemisphere, and the reverse. (I can still remember a poster in my classroom when I was five that showed Australians having their Christmas dinner as a beach barbeque!)
3. That in some parts of the world (near the equator) day and night lengths stay much the same all the year round and that they do not have seasons as in the UK or Australia. (I used to discuss it always getting dark round about six o'clock when I worked in Kenya.)

The Moon

Here the main things to help the children observe are that:

1 The Moon is not always visible.
2 Sometimes it is visible at night but, at other times, during the day.
3 That it appears to change shape and what the names of these shapes are.
4 To reinforce the concept, discussed in Chapter 12, that the Moon does not actually make light but rather reflects the light of the Sun.

The last point can be harder for the children to appreciate as the children will naturally often be puzzled at how the Moon can be lit up by the Sun, even when they cannot see the Sun itself. This presents scope for some simple modelling. I get the children to work in threes in a darkened room. One child is the Earth; the second stands in front of them and holds up a polystyrene sphere, or something similar, to represent the Moon; the third child stands behind the 'Earth' with a torch to represent the Sun. If the torch is shone on the sphere, the children quickly see how the Moon can be lit up but yet they cannot see the Sun. If you are working with more able children, this can also represent a chance to introduce the fact that the Earth is roughly spherical and that it rotates on its axis and even that the Moon orbits the Earth. This helps the children understand why we can't always see the Moon; we can only see it when we are facing it. It also introduces the concept of why we have day and night. Traditionally, these last points tend to be considered key stage 2 work, and it is true that I would not initially plan to cover them earlier, but I have found that the concepts often arise when curious younger children are discussing these areas and so it is best to be prepared to let the children explore them.

Hint!

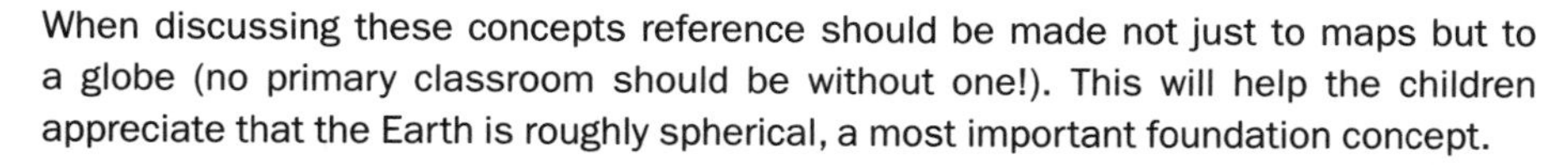
When discussing these concepts reference should be made not just to maps but to a globe (no primary classroom should be without one!). This will help the children appreciate that the Earth is roughly spherical, a most important foundation concept.

Conceptual step two – understanding day and night

This is the most straightforward of the key concepts for the topic and, with some careful practical work, there are very few children who will have any difficulty with the concepts involved.

Eliciting ideas

In the context of designing a science fiction planet, I did this by asking the children did they think all planets had day and night. This then gave me ample opportunity to hear the children's ideas on what causes day and night.

Be warned!

Most children readily appreciate the fact that it is day when the Sun is up, and night when it is not. The major problem lies in their likely explanation as to why this is. All the evidence the children encounter in daily life suggests that the Sun rises, moves up and across the sky and then moves down and sets. On the other hand, there is no obvious evidence to suggest to them that the Earth is spinning round. It is hardly surprising then, if they reach the same conclusion as many of their forbears and believe that it is the Sun that is orbiting the Earth. If the children do suggest that day and night are caused by the Sun orbiting the Earth, it is useful to ask them can they think of any other ways that the same effect might be caused. The various ideas can then be considered and evaluated.

I have also found that children regularly offer metaphysical explanations for phenomena connected with the topic. For example, children have told me things such as, 'we have day and night because God made it that way'. Interestingly, with the exception of some discussions about creationism, I have never heard children advance such suggestions for any other topic in science. Such explanations may seem outside typical science curricula but they raise important issues that must be carefully addressed.

Challenging the children's ideas

I have sometimes seen it suggested that a good way to show the children that the Earth is revolving on its axis is to do a shadow investigation, where the children investigate how the length and position of shadows change during a sunny day. This is a good investigation and helps the children in their understanding of shadows but the results they will record could be explained equally well if it was the Sun that orbited the Earth every twenty-four hours! Instead other challenges must be used.

In fact, I have never found the children having difficulty accepting that it is the Earth that spins, so long as they are given a chance to practically experience the concepts. I normally employ three steps:

1 In pairs, in a darkened room, having one child hold a torch representing the Sun and the other child representing a child standing on the Earth. They begin by facing the Sun, so their face is in the light. Then they slowly spin round on the spot (the Earth spins anticlockwise, if viewed from above the North Pole). As the child moves out of the light and into the dark, it is night time; then as they move back into the light, it is day time once more.
2 Simulating the above but with an un-shaded light bulb representing the Sun and a globe for the Earth. It is useful to place a marker on the position of the children's school. As the globe spins, they will see their school move in and out of the light. Half the globe will always be in shadow (night) and half in light (day).
3 A discussion of the fact that astronauts have now gone into space and watched the Earth spinning accompanied by video evidence of the Earth rotating and computer simulations of this.

Metaphysical explanations such as, 'God made it like this' require careful consideration but I have always found children to be both interested and thoughtful in such discussions. Ideally I hope that, through discussion, the children will come to appreciate that there is a significant difference between statements such as:

- we have day and night because the Earth spins on its axis once every twenty-four hours;
- we have day and night because the Sun orbits the Earth once every twenty-four hours;
- we have day and night because God made it that way.

The key difference is that the first two statements can be tested, the third cannot. In the slightly more technical terms of scientific philosophy the first two statements are potentially falsifiable. On the other hand, it is impossible to prove the third statement is false; after all, who can say that God didn't make the Earth spin round once every twenty-four hours? For some, the fact that the third statement can't be tested and potentially falsified is its great weakness; for others that is its strength and they will consider this to be the whole point of faith. From the point of view of the children's learning, the key fact is that the statements work in different ways. I always find it useful to give further consideration to this in, for example, religious education, or a philosophy for children lesson.

Conceptual step three – understanding the seasons

This is harder than day and night but, once more, a careful use of practical activities will allow the children to get to grips with the tricky concepts.

Eliciting ideas

Designing planets works very well as an elicitation exercise in this respect. I start by discussing with the children what we mean by seasons and what they know about the seasons on different parts of the Earth. They then consider what sort of seasons they would like for their fictional planet and this leads naturally to a discussion about what causes the seasons and therefore how they would have to design their planet to achieve the seasons they wished.

Be warned!

I have found that children nearly always often advance sensible ideas for what might cause seasonal differences but, of course, just because they are sensible suggestions, doesn't mean they are scientifically correct.

One of the most common problems is that children mix up weather with the seasons, often with elements of circular reasoning. For instance I have regularly heard children advance explanations such as, 'it's always cloudier in the winter and that's why it's colder'. Of course, the two are interrelated but weather depends on much more than the season and, although we may associate different weather with different seasons, the weather is not the cause of the season.

One of the most common misconceptions regarding the seasons illustrates the old adage about 'a little knowledge being a dangerous thing!' Frequently children have told me that 'we have winter when we are far away from the Sun but when we're close to the Sun, it's summer'. These children are often aware that the Earth's orbit around the Sun is elliptical and so this reasoning is quite sophisticated and, at first sight, very plausible. Sometimes, as shown in Figure 14.1, the children's reasoning can be even more sophisticated.

A: The child initially drew this diagram and said, 'it's winter now because the Earth is further from the Sun.' They then realized that according to their diagram there would be two winters and two summers. After some more thought they drew a second diagram, as at B.

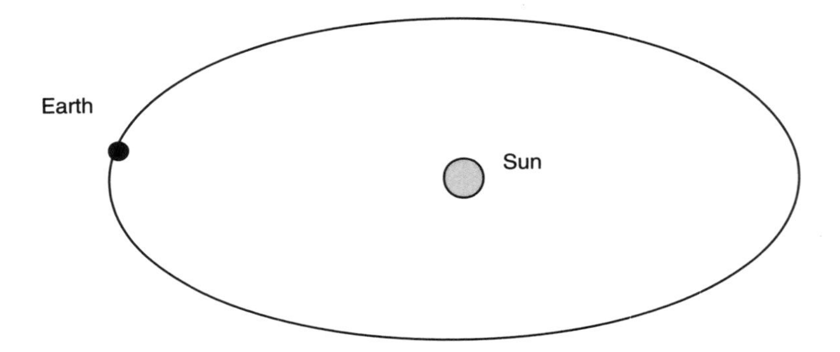

B: The child then said, 'now it's right! The Earth is only close once and far once. It's close now, so it's summer.' Their misconception was challenged by reminding them that when it is summer in the northern hemisphere, it is winter in the southern hemisphere. They realized that, according to their theory, it should be summer, or winter, everywhere on the Earth at the same time and so they needed to think again!

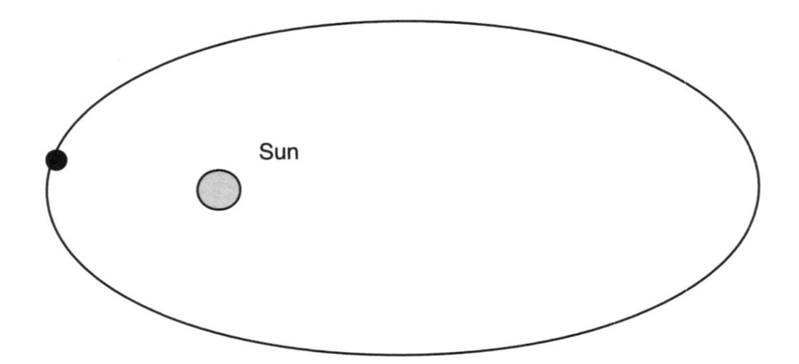

Figure 14.1 Diagrams showing a child's explanation for the seasons

Challenging the children's ideas

There are two points that are vital in challenging the common misconceptions. Both of these points were mentioned in the key stage 1 background.

The first is to emphasize that the key factor in seasonality is varying day length. This helps divert attention from climatic distractions; changing weather can't be influencing the day length.

The other crucial point is that the seasons are not the same everywhere on the Earth at the same time and especially that the pattern is opposite in the northern and southern hemispheres; this means that elliptical orbit theories must be insufficient as, according to them, the seasons would be the same on all parts of the Earth.

In fact, in my experience, children very rarely work out the true cause of the seasons, the Earth's axial tilt. This is hardly surprising, as it is an obscure and difficult concept. Indeed many children may not even realize that the reason model globes are, as one child put it to me, 'slanty', is because that is how the Earth actually is; its axis being tipped at an angle of 23.5 degrees. I am happy to confess to my students that, when I was a rather orderly primary one child, I was irritated by the fact that the classroom globe was 'slanty' and that my best explanation for this was the somewhat nationalistic one that it was so that we could see our country better. (I suspected that South Africans had their globes the other way up!) So, the first step is to draw the axial tilt to the children's attention. After doing this, if the children are encouraged to 'orbit' a globe around an unshaded lamp, some will spot that sometimes the northern hemisphere is tilted towards the Sun and gets more light (at which point the southern hemisphere is tilted away from the Sun) and that, at the opposite point of the orbit, the northern hemisphere is tilted away from the Sun and gets less light (at which point the southern hemisphere is tilted towards the Sun). In the in-between phases of the orbit, the hemispheres are tilted neither away from, nor towards the Sun. It is likely, however, that you will find that you have to help many children notice this and to explain its significance.

Be warned!

When either demonstrating this yourself, or letting the children explore the activity, you must ensure that the direction of the Earth's axial tilt stays the same; the Earth wobbles a little but it doesn't swing wildly about! The easiest way to do this is to always keep the bar of the globe's axle pointing the same way.

Even though the scale is hopelessly out, this activity demonstrates the principles very well. In other words:

1 When a hemisphere is tipped towards the Sun, it is summer in that hemisphere.
2 The days will be longer and the nights shorter. (The children can see this by putting a marker where their school is and rotating the globe on its axis.)

3 The Sun's rays will be stronger because they hit the Earth closer to the perpendicular; therefore the heat particles (Chapter 10) and light particles (Chapter 13) have less atmosphere to get through and so less chance of being absorbed, or reflected. (Once more, this is a simplification but it makes the concepts accessible to the children without getting into the much too difficult area of energy.)
4 When a hemisphere is tipped away from the Sun, it is winter in that hemisphere and the opposites of points 2 and 3 apply.
5 The further north or south a point is, the more extreme the effects. For example in the summer at the North Pole, the Sun never sets; but in the winter, the Sun never rises. (It is possible to show this using the globe activity but it requires some careful experimenting to get the angles and distances correct.)
6 When the hemispheres are tilted neither towards nor away from the Sun, it is spring in one hemisphere and autumn in the other, depending on which point of its orbit the Earth is at.
7 For locations near to the equator, there is very little seasonal variation as they remain always at more or less the same angle to the Sun.

I find that, even with the model, it can take the children quite a while to grasp this. Don't rush them and give them plenty of opportunities to try the model themselves. Computer simulations may help, as might diagrams, but I have found these more useful in consolidating the concepts than in helping the children initially grasp them.

After the children are secure in the concepts it is useful to spend some time discussing other influences on the weather as well as the effects of day length. For example, altitude is an obvious factor, which the children readily grasp. A more complex example is fact that the UK is at much the same latitude as much of Russia, yet the winters in Russia are much colder than in the UK. This is due to the effect of the UK being so close to the sea and especially the warm current of the Gulf Stream.

Conceptual step four – understanding the phases of the Moon

This can be a difficult concept for the children to grasp. In fact I don't think it is as difficult as the seasons, so long as the one effective practical activity is used. If it is not, and especially if some of the inappropriate activities that are sadly commonly suggested are employed, there can be some very confused children indeed!

Eliciting ideas

When discussing the children's fictional planets, the subject of moons soon came up. And, without any prompting, the discussion moved to the phases of the Moon, with some of the children wondering if all planets' moons 'change shape'. This then allowed questioning as to why the children thought our Moon 'changed shape'.

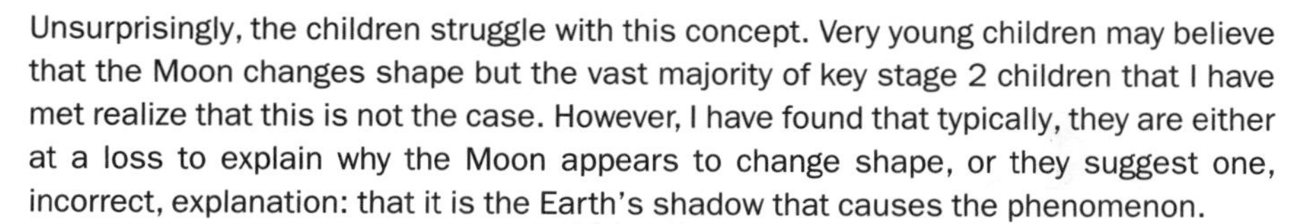

Be warned!

Unsurprisingly, the children struggle with this concept. Very young children may believe that the Moon changes shape but the vast majority of key stage 2 children that I have met realize that this is not the case. However, I have found that typically, they are either at a loss to explain why the Moon appears to change shape, or they suggest one, incorrect, explanation: that it is the Earth's shadow that causes the phenomenon.

Challenging the children's ideas

One of the first things that it is helpful for the children to do is to observe what the pattern of the changing phases of the Moon is. It is best to do this from nature but, depending on the hours the Moon is above the horizon, and on the weather, this is likely to be only possible to a limited degree. At any rate, with as much direct observation as possible and then by using secondary sources, it is important to establish what the pattern of the phases is. After this it is easy to challenge the idea that the phases are caused by the Earth's shadow. I let the children work in teams in a darkened room, giving each team a torch to represent the Sun, and two polystyrene spheres on sticks to represent the Earth and the Moon. Their challenge is then to replicate the phases by using the 'Earth' to cast a shadow on the 'Moon'. They will be able to do this for some phases: full moon (no shadow at all); crescent moon; and new moon (in complete shadow). However, they will soon realize that it is impossible with the curved shadow cast by a sphere to replicate a half moon or a gibbous moon.

This activity shows the children that it cannot be the Earth's shadow that is responsible for the phases of the Moon but, just as with the seasons, they are likely to need some prompting if they are to come up with the actual explanation.

The best practical activity for demonstrating the phases of the Moon is actually quite straightforward, though only one child at a time can take part in such a fashion so as to see the modelled phases. Once again a darkened room is necessary and a bright light to represent the Sun. An unshaded lamp set on a high table will do, so long as it is a bright one. (Overhead projector lamps used to be ideal, but, in an age of data projectors, they may be hard to find.) Alternatively, a second child could stand to one side and hold a torch to represent the Sun. The modelling child represents the Earth and stands to one side of the lamp, holding a polystyrene sphere on a stick to represent the Moon (my children have always called this a 'moonstick'). The child holds out the 'moonstick' and faces it. They need to hold the 'Moon' slightly above their head, so their shadow doesn't interfere with the phases (otherwise they would model a lunar eclipse!). They then slowly rotate anti-clockwise, facing the Moon all the time. This represents the Moon orbiting around the Earth. This model leaves out the fact that the Earth should have rotated roughly 29.5 times for each orbit of the Moon but attempting to synchronize that is very difficult! It is important that the children realize this, however. As the Moon orbits the child, they will see that different portions of it are lit up at different times, whilst other portions remain in shadow. This will replicate perfectly the phases of the Moon. Figure 14.2 shows how the model works using as

Plan view of a child holding a 'moonstick'. As we are looking from above, we see the moon as half lit, half in shadow, but this is not the view the child sees, which is illustrated below.

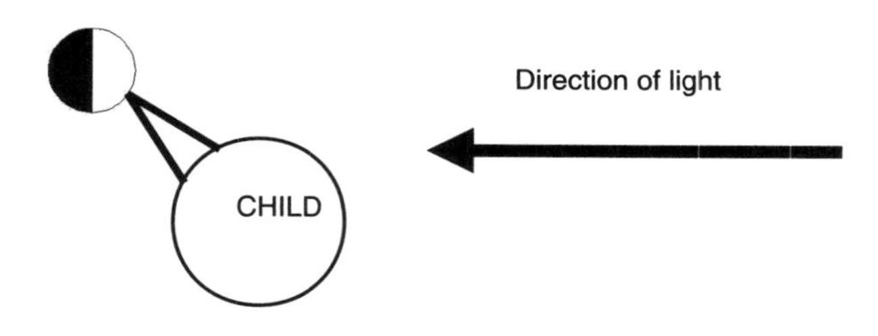

The view of a 'moonstick' a child would have in this position; a waxing gibbous moon.

Figure 14.2 Using a 'moonstick' to model the phases of the moon

example the view a child would have of a gibbous Moon. It is actually a waxing gibbous Moon, as the Moon is moving towards the full, where all of it that is visible to the child will be illuminated.

This is the only practical activity that I have ever found to be of any use when helping children learn this concept. Many of the others that I have seen suggested that involve footballs painted half black, or squinting into boxes, are not only confusing for children but, by portraying mechanisms somewhat different to what actually happens, are only likely to foster misconceptions.

Another advantage of this activity is that it shows how the Moon is lit up even when the Sun may not be visible to us. As was mentioned in Chapter 12, this can confuse children, who sometimes find it difficult to accept that the Moon is not a light source and is only reflecting the sunlight.

Another source of confusion that I have sometimes encountered is when children have heard of what is termed 'the dark side of the Moon'. The activity demonstrates that there is no dark side of the Moon; only half of the Moon at a time is ever illuminated but, at some point or other, every part of the Moon will have light falling on it. The strange fact is that, despite the fact that the Moon is spinning on its axis, we only ever see the same face of it. The face we never see is called 'the dark side', even if it not necessarily dark. This comes about because of the amazing coincidence of the Moon taking almost exactly the same time to rotate on its axis, as it takes to orbit the Earth. You can try demonstrating this to the children with a marker on one face of a 'moonstick' and you orbiting it round them, whilst at the same time carefully synchronizing its rotation but it takes practice!

15

Developing a primary school model of sound

Introduction

Sound is by definition a noisy topic and perhaps then it is no wonder that children usually find it great fun. It is not so conceptually difficult as the other physics-based topics such as light, electricity, or forces but it still has problems for the unwary.

Where this topic fits in

The first two conceptual steps can be covered at key stage 1 and are straightforward enough to be addressed even early in the key stage. The final conceptual step is best covered in key stage 2, though, once again, it is not especially complex and can be covered early in the key stage.

Links to other topics

The concepts addressed here are not vital for the children's understanding of other science topics but, once again, a good grounding in the concepts of materials and their properties (covered in Chapters 6–9) is essential if the children are to make sense of the phenomena they encounter and build an appropriate model of sound.

Depending on the learning context chosen, there could be many good cross-curricular links for the topic. However, the opportunity for excellent cross-curricular links with music is really too good to pass up on. Particularly useful links with technology may also be made since making musical instruments is, especially at key stage 2, one of the very best ways for children to learn about the concepts of sound.

A context for the topic

I have found that by far and away the best context for sound is music making. This allows for all the key concepts to be addressed with plenty of emphasis on science skills. For the younger children, most of the emphasis might be around working with ready-made instruments along with making some simple percussion instruments.

Table 15.1 Conceptual steps and example learning objectives for developing a primary school model of sound (conceptual steps one and two refer to work best undertaken in key stage 1; the final conceptual step is best covered in key stage 2)

Conceptual step	*Example learning objectives*
	Children will learn:
Identifying sounds.	to identify a wide range of sounds; to be able to describe sounds using appropriate scientific vocabulary;
Understanding a simple key stage 1 model of sound.	that there is always movement associated with sound; that sounds travel away from their source, getting fainter as they do so; that we detect sounds using our ears;
Refining and extending the basic model of sound.	understand that sounds are produced when materials vibrate; to distinguish between, and to control, loudness and pitch; understand that when sound travels, it must be via a vibrating material.

Older children can be much more ambitious and carry out the bulk of their learning through making their own instruments. As you can imagine, children respond very well to this and I have always found the topic an excellent springboard for both music skills and technology, as the children design and make their own musical instruments.

Conceptual step one – identifying sounds

At key stage 1 this is largely an exercise in developing the appropriate scientific vocabulary.

Eliciting ideas

Given a music making context, it is easy to pre-assess the children's ability to describe and categorize sounds by getting them to describe the different sounds made by various instruments. It is particularly important to ask them how they can vary the sounds made by different instruments, as this will lead into discussions of loudness and pitch.

Be warned!

This activity will raise a wide range of general vocabulary, which will aid the children's language development but it is the scientific vocabulary that is our primary interest here and this relates to two key areas: pitch and loudness.

Distinguishing between loud and quiet sounds is unlikely to present problems for the children, though you may have to take care over potentially confusing vocabulary such as 'soft' being used to describe quiet sounds.

Distinguishing between pitch may be trickier. Some curricula suggest that pitch should not be introduced to younger children but it will form an integral part of their music and, when thought of in terms of high and low notes, I have never found it too taxing a concept for even younger children. Sometimes a child may get confused between pitch and loudness but I have actually found this more prevalent amongst key stage 2 children who have not had a good practical grounding in describing sounds earlier in their learning.

Challenging the children's ideas

The aim here is to simply ensure the children get lots of practical opportunities to make, listen to and describe a wide range of sounds. Even regarding distinguishing between high and low notes I have never found the children to have great difficulties, although, as is always the case with younger children, it is best to, at least to begin with, ensure that there is initially a clear distinction between the phenomena they have to distinguish between; make sure the high notes are clearly higher than the low ones!

Conceptual step two – understanding a simple key stage 1 model of sound

As always, it is most helpful to assist the children in building up a model to explain the phenomena they encounter and investigate.

Eliciting ideas

Again, the musical instruments offer a good lead in to the discussion. I ask the children how they think the sounds are being made and how they know there are sounds being made. The latter question will certainly show that the children are aware that they use their ears to hear but it is interesting to follow this up by asking them what they think the sounds actually are and how they get to their ears.

Be warned!

I have found that children often find it difficult to link sound and vibration but, especially if the topic is being tackled early in key stage 1, I am not too concerned with this. However, it is important to be aware that it is likely, at this stage, that the children's understanding will be incomplete. Many will simply attribute sound to an action rather than understand that the action is what causes the vibrations. They will make statements such as, 'the drum makes a sound because I hit it', but not be aware that it is the drum vibrating after it is hit that causes the sound.

Even young children are usually well aware that they use their ears to hear but some think that hearing is an active process, in a similar fashion to their thinking that sight is an active process. Just as with light, the confusion is understandable and is often reinforced by everyday speech such as their being told to 'listen carefully'.

When it comes to suggesting what sound actually is, I have found that young children are usually rather baffled. They frame ideas such as by describing sounds and I have often heard statements such as, 'sounds are made by noises', or 'sounds go away from the loud place and sort of run out'. So, once again there is a need for a simple model to help them understand their observations.

Challenging the children's ideas

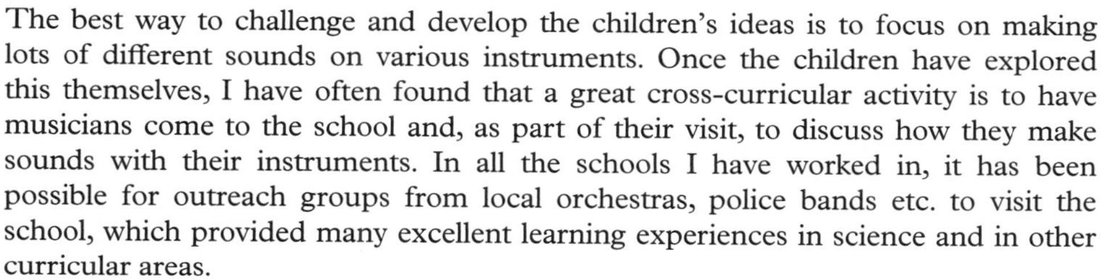

The best way to challenge and develop the children's ideas is to focus on making lots of different sounds on various instruments. Once the children have explored this themselves, I have often found that a great cross-curricular activity is to have musicians come to the school and, as part of their visit, to discuss how they make sounds with their instruments. In all the schools I have worked in, it has been possible for outreach groups from local orchestras, police bands etc. to visit the school, which provided many excellent learning experiences in science and in other curricular areas.

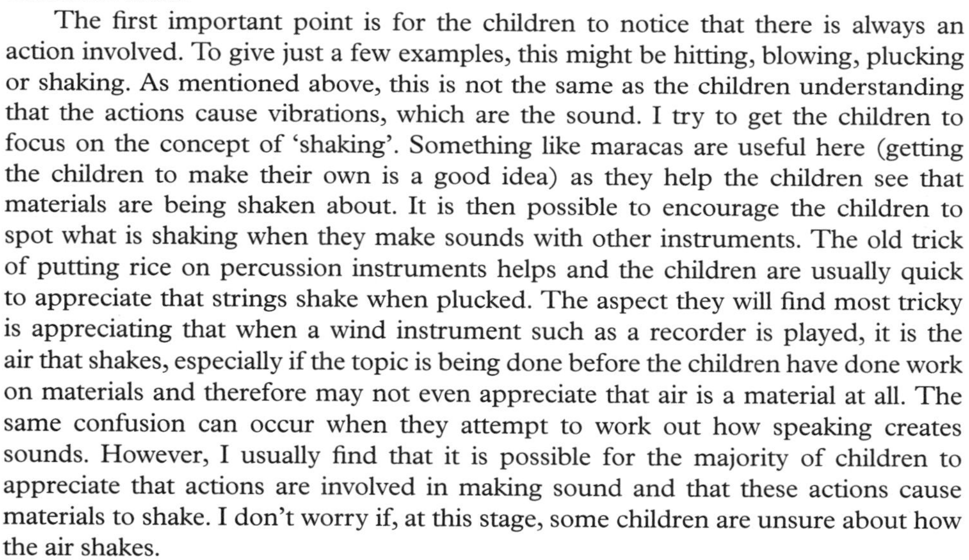

The first important point is for the children to notice that there is always an action involved. To give just a few examples, this might be hitting, blowing, plucking or shaking. As mentioned above, this is not the same as the children understanding that the actions cause vibrations, which are the sound. I try to get the children to focus on the concept of 'shaking'. Something like maracas are useful here (getting the children to make their own is a good idea) as they help the children see that materials are being shaken about. It is then possible to encourage the children to spot what is shaking when they make sounds with other instruments. The old trick of putting rice on percussion instruments helps and the children are usually quick to appreciate that strings shake when plucked. The aspect they will find most tricky is appreciating that when a wind instrument such as a recorder is played, it is the air that shakes, especially if the topic is being done before the children have done work on materials and therefore may not even appreciate that air is a material at all. The same confusion can occur when they attempt to work out how speaking creates sounds. However, I usually find that it is possible for the majority of children to appreciate that actions are involved in making sound and that these actions cause materials to shake. I don't worry if, at this stage, some children are unsure about how the air shakes.

The latter point can make it difficult for young children to appreciate how sounds travel to their ears. Once again, this is not crucial at this stage, some children will appreciate that sound travels through materials by shaking them but I concentrate on helping the children see that sounds travel in all directions from a source and get fainter as they do so.

The notion of active hearing can be tricky and is not so easily challenged as the misconception regarding active sight. Even so, I find that most children, having spent time on the concept that sounds are produced when materials are caused to shake, are happy to accept that it is the sounds that travel away from the source and that we hear them simply when they go into our ears.

Conceptual step three – refining and extending the basic model of sound

The essential foundations have been laid in key stage 1 and it is now a matter of refining the children's understanding of how exactly sounds are made and travel.

Eliciting ideas

Music makes as good a context for sound in key stage 1 as it does in key stage 2 and it is rarely that I would not have the children make their own musical instruments, but it is possible to vary the musical context a bit and set the children some investigative science tasks as a great way to elicit their ideas.

I use many puppets and characters and I have always been delighted with how even older children respond to them. For younger key stage 2 children one of my most successful helpers has been Scrumpsy, the small beige dog, a friend of Powell the owl (whom we met in Chapter 6). In fact Powell often roosts on top of Scrumpsy's house, which can cause problems because Powell is usually settling down to sleep just as Scrumpsy is getting up. Scrumpsy is a regular jazz puppy and his alarm clock wakes him by loudly playing Art Blakey. The problem is that this often wakes Powell up and the two friends want to find a way of stopping the sound disturbing Powell but still being able to wake the soundly sleeping Scrumpsy. The children soon notice there is no door on Scrumpsy's house and they easily focus their efforts on trying to design a soundproof one. This leads nicely to great investigative science using a computer data logger to measure the sound whilst testing a range of different door types. The resulting discussions are a good way of finding out the children's ideas on how sound is made, how it travels and its reflection and absorption.

Be warned!

Discussions such as these may reveal several problems.

For example, I have found that, although the majority of children readily accept that sound travels through air, some think it does not travel through other materials.

You are also likely to find that the children's understanding of the fact that sound can be reflected and absorbed is at best partial. Reflection causes less problems and, indeed, you may find that some children are familiar with this through experiencing echoes. Absorption, on the other hand, causes more difficulties and, even though some children will accept the fact that putting a door on Scrumpsy's house will help keep the sound in, they may think this is because the door bounces all the sound back, rather than both reflecting it and absorbing it.

In the discussion around the last conceptual step, it was highlighted that children may not realize that sound is vibration and instead think vibrations simply cause sounds or the reverse. This is still likely to cause difficulties, especially if the vibrations are far from evident, as, for example, in the case of the speaker of Scrumpsy's alarm.

Figure 15.1 Young science investigators help Scrumpsy soundproof his house

Challenging the children's ideas

The investigation provides many ways of challenging the children's ideas and makes a good foundation for building up a model of what sound is and how it travels.

It is usually best to begin with a revision discussion of how sounds are made. In the previous conceptual step, it was emphasized that children should be able to associate sound with some sort of movement e.g. plucking a string or hitting a drum, but that at that stage, some would not appreciate that the sound is the vibrations that are caused by the plucking, hitting etc. It is important to revisit these ideas now and, once again, music provides a good lead in, for example, Scrumpsy often asks the children how they think the instruments in the jazz band make sounds. The strategies discussed to emphasize the vibrations can be revisited and applied to less obvious contexts. For example, a sheet of paper can be laid over a speaker and some fine play sand, or similar, laid on it. At high volume and with a strong bass line, the children will be able to see the sand jump, revealing the vibratory nature of sound. Other less obvious examples include blowing on a blade of grass held between your two thumbs to make it whistle and, in a similar fashion, making whistles from art straws. I have also found it important to emphasize examples where it is only the air that vibrates, such as blowing across a plastic bottle top, or playing the recorder. This is important as, as is often the case, the children can find concepts associated with seemingly incorporeal gases trickier to grasp. The key point in these discussions is to let the children see that sounds are vibrations in materials as the particles squeeze up and stretch out.

This leads naturally on to discussing how sounds travel, which is easy to discuss in the context of the investigation; if the data logger detects a noise, sound must be travelling out from the speaker through the various materials between it and the sensor. It does this by the particles squeezing and stretching as the vibration passes through the material. It is important that the children appreciate that this can happen through any material. The investigation facilitates measuring sounds that have travelled through a gas (the air) or solids (various doors) but you must also allow the children to detect sounds that have travelled through liquids, for instance by tapping inside a tank of water. This then leads naturally to a discussion of what sounds might be unable to travel through; in other words a vacuum, where there are no particles to vibrate.

This then poses a problem for the children in that, if sound can travel through any material, how can they allow Powell to sleep peacefully by putting a door on Scrumpsy's house? In fact, I find that the children typically quickly appreciate that some of the sound vibrations are being bounced back. (They can also model this with a 'slinky' spring and, of course, link it to echoes.) They do often find the concept of some vibrations being absorbed harder but, in essence, all they need appreciate is that some materials are more difficult for vibrations to travel through than others; the whole point of the investigation, as they investigate different materials and combinations of materials.

Finally it is necessary to return to the difference between volume and pitch. After the previous conceptual step, the children should have an awareness of the differences between volume and pitch. It is now possible for the children to extend their model to explain the differences.

I find that the children quickly appreciate that the reason some sounds are louder than others is because the particles in the materials are vibrating more. There is a straightforward conceptual link here: hit the drum harder and make the particles vibrate more; therefore a louder sound. The same logic applies to blowing a whistle harder, or plucking a string more forcefully.

Pitch can be harder. The notion of the particles vibrating at different frequencies can be rather abstract for the younger key stage 2 children, for whom this topic is ideal, and it is not essential that they realize this. Accordingly, I concentrate on their being able to link changes in pitch to characteristics they can easily observe, for example, strings being under more tension, or changing the length of an art straw whistle. Here there are some great opportunities for the links to music and to design and technology mentioned earlier, as the children can design their own instruments. Perhaps they can even have a go at their own jazz improvisations with Scrumpsy, or a similar friend!

Be warned!

We routinely talk of sound waves and this is correct; sound does travel in waves. However, I have found that this term often confuses the children who think of waves as only being the 'up-and-down' longitudinal waves, for example those formed by dropping a stone into water. The 'squeezes-and-stretches' of compression waves, such as sound waves, are much harder for the children to appreciate. Squeezing and stretching a 'slinky' spring can help the children appreciate this concept as the links of the spring will bunch in squeezes and stretches, just like the particles in a material as a vibration moves through it. It may well also be helpful to avoid the term 'wave' at all and I certainly place the emphasis on the term vibration.

16

Developing an initial understanding of what goes on in a simple circuit

Introduction

Electricity is a topic that primary school children tend to enjoy; they have a lot of fun with the gadgetry used to construct circuits and relish the practical tasks this involves. It was one of the first science topics that I ever had to teach and I enjoyed it as much as the children did. However, it is sobering to look back on that early work as I now realize that, fun or not, I was not doing a very effective job at helping the children learn and, worse still, I was probably confirming the children in some common misconceptions. I now realize that electricity is undoubtedly one of the hardest primary school science topics to teach, the more so because it can appear deceptively straightforward. The common misconceptions regarding this topic are particularly insidious and so we have to work hard to identify them and challenge them, remembering that, just because the children can build working circuits, doesn't mean to say they understand them. Even so, if we draw on the valuable research that has been done on children's learning in this area and plan carefully, there should be nothing to fear. This chapter is designed to alert you to the key pitfalls of the topic and provide guidance on how to progress children's understanding of simple circuits.

Be warned!

Often an open-ended, exploratory approach is a great way to elicit the children's ideas but don't try this with electricity! If you do, curious children will quickly discover phenomena that will leave them completely out of their depth to explain. The ideas in this chapter will help you follow a structured approach that will avoid such problems, yet can still centre on the children's ideas.

Where this topic fits in

There is a broad spread of conceptual progression in the topic and it is very important that children get a strong foundation in key stage 1. However some of the concepts are rather abstract and tricky, even at this stage, and so, where possible, I try to use the topic later in the key stage, when the children's skills will have advanced further.

It is very useful if the children have covered the aspects of materials and their properties discussed in Chapters 6 and 7 before tackling electricity as this will help them explain some of their observations that they might otherwise struggle with.

A context for the topic

Electricity is unusual as it is the only primary science topic that I teach as a stand-alone unit and without being set in a broader learning context as has been outlined for the other topics in the book. I did not always follow this approach. When I started teaching I tried combining learning about electricity with design and technology projects, an obvious and seemingly useful link, but I soon realized this caused problems. The children became absorbed in the technology problems and the less structured approach caused problems with learning about electricity. I now recommend learning about electricity is addressed in stand-alone topics, which allow for much more precise focus on the difficult conceptual aspects of this topic. I don't find this a problem as the children are excited and enthusiastic about building circuits anyway. However, I do incorporate cross-curricular links and more open-ended, exploratory learning after the initial work. That is the time I use all those great design and technology projects such as powering buggies and constructing burglar alarms. Once the main conceptual progress for electricity has been addressed, such follow-up activities are a splendid means for the children to apply and embed their learning. Such activities also provide excellent opportunities for you to assess the children's understanding of the scientific concepts.

Choosing equipment

This topic is one where choice of equipment is vital, not just for reasons of practicality or safety, but because poor choices can seriously hamper the children's conceptual progress. Advice on this is given in a section at the end of Chapter 17 that covers what equipment to use at both key stages. The advice given there will help you avoid many pitfalls and help prevent the misconceptions that the children may develop if presented with poorly chosen equipment.

A note regarding electricity and science skills

The introductory sections of the book stressed how combining the development of the children's science skills alongside that of their subject knowledge and understanding would assist the children to connect up with their own ideas and so help their learning. This is true for the topic of electricity but when drawing up your science programme, it is worth remembering that electricity as a topic is not a suitable one for supporting science investigations. (The main difficulty is that, at primary school level, it is impossible for the children to effectively measure the likely dependent variables.)

However, the topic does support the development of certain skills very well and I find the most useful ones to concentrate on are:

- predicting;
- using knowledge and understanding to explain results;
- using equipment appropriately;
- presenting information.

Conceptual step one – introducing electricity at an everyday level

When introducing the topic to young children it is important to check their ideas about electricity at the most basic level and establish some basic foundation concepts before going on to introduce simple circuits.

Eliciting ideas

At this stage the children are only being asked to think about how they encounter electricity on a daily basis and I concentrate on checking the children's ideas in two areas, both of which can be introduced as simple discussion topics. The two areas to examine are: where the children think electricity comes from; and what it does/how it is used.

Table 16.1 Conceptual steps and example learning objectives for developing an initial understanding of what goes on in a simple circuit (these learning objectives refer to the introduction of concepts in key stage 1)

Conceptual step	*Example learning objectives*
	Children will:
Introducing electricity at an everyday level.	understand the distinction between mains and 'battery' electricity; be able to identify appliances in the home/school that utilize electricity; understand the dangers of mains electricity and know and apply appropriate safety measures;
A basic understanding of circuits.	be able to draw their circuits and to construct simple circuits from drawings; understand that a complete circuit is required to light a bulb (or work other devices) and why this is so; know the functions of battery, wires, bulb and buzzer and be able to explain these in appropriate terms; be able to build simple series circuits to solve basic technological problems;
Introducing switches.	understand how switches work; be able to construct a simple switch; be able to incorporate switches in their simple circuits.

Hint!

If you wish to discuss these matters with smaller groups of children a useful tactic is to set the class a simple cutting and pasting exercise, which will also provide you with useful pre-assessment information. For instance, the children could be given old catalogues and asked to cut out pictures of things that work by electricity. This keeps everyone gainfully occupied while you can go round and discuss the children's ideas with groups of children rather than the whole class.

Be warned!

The children can usually easily identify lots of appliances that are powered by electricity and state what they do. They are also usually quick to spot that the appliances can be powered by batteries or by the mains (although, for the latter, they are more likely to use a statement such as, 'it works by a plug'. However they are likely to be much less clear on what exactly electricity is and how it actually works the appliances.

Challenging the children's ideas

As has been said, the children quickly appreciate that electricity can be provided to appliances through either the mains or from batteries. You are unlikely to have to challenge misconceptions in this area and simple activities will help the children discuss and embed the concepts such as:

1 Making a collection of appliances that use batteries;
2 Making 'electricity maps' of rooms in their homes showing the appliances that use electricity;
3 Identifying the appliances in the classroom that use electricity by putting stickers on them;
4 Cutting out pictures from old catalogues to make sets of appliances that use mains electricity, use batteries, and do not use electricity.

At this level, it is not at all appropriate to attempt to explain the differences between mains and battery electricity (indeed this is completely beyond primary school level) but it is important that the children appreciate the dangers of mains electricity. I have found that it is best to encourage the children to limit their explanation of the difference to concepts such as, 'mains electricity is more 'powerful' and so more dangerous than the electricity from batteries. Once again, the learning here is very straightforward and can be addressed through simple activities such as:

1 Identifying hazards in pictures;
2 Producing safety posters;
3 Writing rules for electrical safety in the classroom/home.

Regarding the more complex concepts, I do not at this stage attempt to challenge any misconceptions that may surface regarding what electricity actually is, preferring to tackle this when the children have done some more work on building circuits, as described under the next conceptual step. The children will also have all kinds of ideas about what the electricity does to make appliances work but at this stage it is not appropriate to get into detail. I have found that it is better to help the children suggest simple formulae that highlight what it is that the electricity actually enables, for example:

- electricity works the *light bulb* to *give us light*;
- electricity works the *lawn mower* to *make the blades go round and cut the grass*;
- or electricity works the *oven* to *make it hot and cook the food*.

Conceptual step two – a basic understanding of circuits

This is a much more complex area than the previous one, despite the fact that the learning objectives appear deceptively straightforward. The scientific explanations involved are often counter-intuitive and matters are made more difficult by the fact that the children cannot actually see what is going on in a circuit, only some of its effects. It is hardly surprising then that the children have many misconceptions in this area and that challenging them is a priority if the children are to gain a solid conceptual foundation before going on to more complex concepts.

Eliciting ideas

One of the best ways to elicit the children's ideas in this area is to give them some simple equipment, ask them to see if they can get the bulb to light, and question them about their ideas as to what is going on as they explore the task. When questioning, it is important to probe the children's understanding not just of how to connect up the components but also as to the functions of the different components and as to what they think is happening in the circuit. The exercise is best done with the children working in groups no larger than pairs and, since you will wish to observe and question the children, you may wish for only sections of the class to undertake this activity at a time. As has been mentioned, choice of equipment is very important and for this exercise I give the children:

- a 1.5V battery* in holder;
- a 2.5V bulb in holder;
- and half-a-dozen connector wires with crocodile clips.

* Technically this is a *cell* not a *battery* but the distinction is not important at primary level.

Most children will manage to get the bulb to light but often after a considerable amount of experimentation, using varying numbers of wires and connecting them in different fashions. You are also likely to encounter a number of imaginative but very unhelpful misconceptions.

Be warned!

There are many common misconceptions regarding the functioning of simple circuits.

Regarding the actual construction of the circuits, it is hardly surprising, given the apparent case of mains-powered appliances, that many children will initially believe that only one wire is required from the battery.

You will find your children are much less sure about what electricity is, although many will confidently advance a range of interesting suggestions. Many view it as a sort of liquid flowing through the wires, often equating it with some sort of fuel like petrol or oil. Other common responses include statements similar to, 'it's like sparks and it's dangerous', or, 'it's sparky stuff that lights lights'. You are also likely to encounter other ideas than these. For example, a few years ago I was observing a student pre-assessing some year one children in this area. One of the children claimed that they had seen electricity and, when asked where, explained that it had been 'in the toilet'. With commendable professionalism, but with some trepidation, the student asked the boy to tell her more. The child explained that, 'it was yellow and green and was behind the radiator'. The boy had noticed an earth wire and was equating 'electricity' with 'wire'. Had the student not noticed this prior to the coming work on basic circuits all manner of confusion might have been engendered.

Particularly common is the misconception that the battery stores electricity and that it gets used up in appliances, leading to the battery going flat.

Another common misconception is that electricity flows from the battery down both wires and meets in the bulb to light it. This misconception is so common, it even has a recognized name, *the clashing currents model*.

Challenging the children's ideas

The first step in beginning to challenge these ideas is for the children to begin by building and testing a wide variety of circuits. This will allow them to make observations that will aid them in developing their understanding of what happens in the circuit. I aim to help the children establish the following four points about what is required for a circuit to work:

1 That a battery is required;
2 That the circuit must be unbroken;
3 That all the components have two 'ends', or 'places for connection' (technically terminals) and that both of these must be connected into the loop;
4 That only certain types of materials are used in the circuit where the electricity flows. (At this stage the focus should be on observing that these are metals.)

An important part of such activities will be some means of representing circuits: both for you to give the children instructions and to allow them to make predictions; and for the children to record their work. At this stage, simply drawing the components is sufficient; there is no need for the children to be able to use circuit symbols until later in key stage 2.

Hint!

Many young children will struggle with drawing the components and so it is useful to have cut out pictures of them that the children can stick on paper and join up with pencilled lines to show the connecting wires.

The next step in challenging ideas is to help the children develop a model to understand what goes on in the circuit. Here we really have to help the children as, given the counter-intuitive nature of electricity and the fact that the children can't actually see what is going on in the circuit, there is no way for them to discover what is happening unaided. This is a tricky task and poor explanations can actually reinforce children's misconceptions. However, you shouldn't shy away from explanation as all this does is to let the children come up with their own imaginative but scientifically incorrect theories.

I have met many teachers who believe it is impossible for key stage 1 children to build an understanding of what happens in a series circuit but in my experience this is simply not the case. The following model was developed through a research project I ran with students and school staff. We found it worked very well at key stage 1 and that it greatly reduced later difficulties at key stage 2.

A key stage 1 model for understanding electricity

It has already been suggested that the first science topic tackled at key stage 1 should be materials. If the children have been introduced to materials, they will be familiar with the fact that materials have all manner of different properties and that they are made up of particles. Now the children can be told that a property of all materials is that they contain enormous numbers of incredibly tiny particles called electrons. Some materials have the property that the electrons in them can move around quite easily and that this is what electricity is; the electrons flowing around.

It is true that the children will only have the haziest ideas of what the electrons are. I find that they cannot really grasp the concepts of either the size of the electrons or their incredible numbers and that they sometimes picture the electrons as some sort of liquid, or, more often, as something like tiny marbles rolling around inside the circuit. At this stage however a more exact model is not required and, since not even particle physicists have the faintest idea of what an electron actually looks like, or even where exactly it might be at any given time, we shouldn't panic if the children have no precise notions about this either. The key thing is that the children get access to a model they can use to explain what they observe in circuits and that provides an alternative model to the sorts of misconceptions outlined above.

The children can then act out what happens in a circuit. I do this as follows:

Step 1

The children represent a loop of wire and stand in a circle holding a loop of rope with pieces of string knotted evenly all along its length. The knots represent the electrons, although, of course, each knot really represents an incredibly large number of electrons.

One child is designated as a bulb with a suitable sign around their neck. They have a crown to place on their head when they are lit.

Will the bulb light? No, there is no battery and hence the electrons aren't moving around the circuit.

Step 2

One child can now be designated as a battery and given a suitable sign. The chain of electrons now flows around the circuit and the bulb lights up. If the circuit is broken (a child steps out of the loop) the electron chain stops moving and the bulb goes off.

There is no need at this stage to labour the fact that the electrons flow from the negative terminal on the battery to the positive but, for the sake of correctness, I make sure that the loop moves in such a fashion relative to the picture of the battery.

The children can then discuss their original ideas about how a circuit worked in the light of the model. It will certainly show why a complete loop is required and why appliances must be connected up at both their terminals; the electrons could not flow round otherwise. It also challenges ideas such as the battery being a store of electricity that gets used up by bulbs etc.; the children can see that the electrons are all around the circuit and that they don't get used up but flow round-and-round the circuit.

To fully understand the model, the children will also need to develop a more complete understanding of how the various components in the circuit function. Once again, they will not be able to discover this through any test they can carry out and so you will have to help them. The definitions I have found helpful at this level are as follows.

Figure 16.1 Initial teacher education students practising the key stage 1 model of what happens in a simple circuit

Battery

This contains chemicals that push the electrons around the circuit. When the chemicals stop working, the battery can no longer push the electrons.

Wires

These allow the electrons to flow around the circuit. Some children may now be able to speculate that metals are good for allowing electrons to flow but other materials such as the plastic covering the wires, do not let electrons flow easily.

Bulb

The children should be encouraged to observe that the bulb has a little wire running through it and to understand that this is the path for the electrons to flow through the bulb and on through the circuit.

The electrons flowing through the bulb light it up. Curious children will want to know why! At this stage it's best to encourage the children to make the link that hot things often glow e.g. cooker rings, electric fire elements. It is then an easy step to realizing that the tiny wire in the bulb, the filament, gets hot and glows, giving off light.

Be warned!

Energy saving light bulbs and fluorescent tubes work in a different way but this is really beyond the children at primary level.

Buzzer

It is to be hoped that the children will have studied sound before they cover electricity (it is a conceptually easier topic). If so, even at key stage 1, they should have noticed that sounds are associated with objects moving (vibrating). The electrons flow through the buzzer making part of it shake (vibrate) which makes a sound. The children may well be able to feel the buzzer vibrating if they hold it.

Be warned!

Battery and bulb holders can confuse the children if they are not encouraged to thoroughly inspect them and see how they form part of the circuit. The children should be able to observe how metal parts of the holders connect up to the two 'ends' (terminals) of the battery or bulb and so make it part of the loop of the circuit. It is also a good idea to help the children make a circuit without the holders (they may not have the dexterity to do this themselves) so that they can see the holders are not a necessary feature but are only designed to make life easier.

Conceptual step three – introducing switches

Now that the children have had a thorough introduction to how circuits work, they usually grasp the concept of switches very quickly.

Eliciting ideas

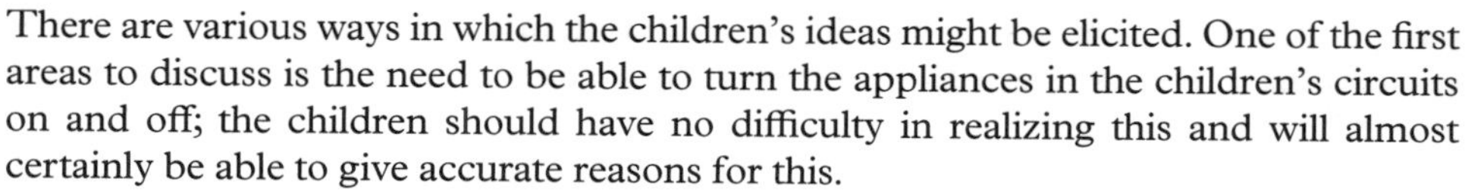

There are various ways in which the children's ideas might be elicited. One of the first areas to discuss is the need to be able to turn the appliances in the children's circuits on and off; the children should have no difficulty in realizing this and will almost certainly be able to give accurate reasons for this.

You can then ask the children how they might turn the appliances in their circuits on and off. Many will suggest, quite correctly, various ways of breaking the circuit, or removing the battery. Some may also suggest using a switch, which will allow the discussion to proceed to how the children think switches work. It may then be helpful to let the children examine examples of commercially produced switches, so long as you choose examples where the children can see the workings of the switch and how it connects and breaks the circuit. In my experience the children are quick to be able to spot how the switches work and they can then go on to making switches of their own.

> ***Be warned!***
>
> At this stage you may well find some children still evidencing misconceptions with statements such as, 'You need to turn the bulb off so as not to waste electricity'. So you may need to revisit the circuit model and definitions of circuit components discussed above.

Challenging the children's ideas

As discussed, you are unlikely to face serious misconceptions concerning switches and the best way for the children to connect with their ideas regarding them is to have a go at building some of their own. Some children may well be able to come up with their own designs, whilst others may require more scaffolding but there are a variety of ways for even young children to construct home-made switches. I find the easiest is to mount two paper fasteners as terminals in thin expanded polystyrene, or cardboard. The paper fasteners can be connected into a circuit with a paperclip hinged on one fastener. The paperclip can then be swung, gate-like, between the two to complete or break a circuit.

Following up these activities

If the children have progressed through these conceptual steps, they should have an excellent grounding in how simple circuits work. The dangers of attempting the above learning through technology projects has already been discussed but now the learning has been tackled, design and technology projects offer great opportunities for the children to consolidate their learning and for you to assess their progress. I try to find projects that fit with cross-curricular learning and the possibilities are practically endless but some examples of projects I have had children engage in include: wiring model rooms for various fictional characters with appliances such as lights, doorbells and fans; building models with working electrical parts of things as diverse as windmills or spacecraft; or constructing simple multiple choice quiz games where a light comes on if a contact is touched to the correct answer's terminal.

17

Developing a more advanced understanding of what goes on in a simple circuit

Introduction

As was discussed in the last section, electricity is a conceptually difficult topic and many of the concepts introduced in this section can be extremely challenging for children to get to grips with. However, if you take care to centre learning on the children's ideas and make use of the advice given here, you should find that matters are not nearly so difficult as they appear.

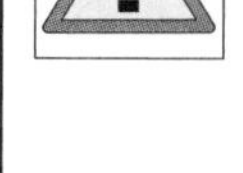

Be warned!

Once again, just as was discussed in Chapter 16, an open-ended, exploratory approach is strongly warned against, due to the likelihood of the children discovering phenomena that will leave them completely out of their depth to explain. The same warnings about combining learning about electricity with design and technology projects also apply. Once again it is a good way to consolidate and assess the children's learning but will prove distracting, should you attempt to use it as a means for challenging the children's ideas and introducing the concepts.

It is also important to remember that choice of equipment is vital and advice is given on this at the end of this chapter.

Where this topic fits in

This topic builds on the concepts introduced in Chapter 16 and it is important that both you and the children are familiar with the concepts covered there before proceeding to those introduced in this section.

Ideally, although given composite classes this may not be possible, I have found that the concepts introduced here be split into two separate slots at key stage 2. There is a very great deal to cover in one unit of work, both in terms of time involved and conceptual progress covered, and splitting the work will help you to manage the learning and the children to build on their previous understanding. I find that covering the first two conceptual steps, *revising the basic principles of simple circuits and representing them schematically*; and *learning about electrical conductors and insulators*; in the first unit, followed by the second two conceptual steps: *a more advanced model for understanding what goes on in simple circuits*; and understanding what happens when the number of components in a circuit is changed; in a second unit, works very well. The first unit can be covered in lower/mid key stage 2, whilst the more complex second unit is best left until upper key stage 2.

It is important to note that for both the proposed units, the children need a good grounding in materials and their properties before tackling the electrical concepts.

Table 17.1 Example learning objectives for developing a more advanced understanding of what goes on in a simple circuit (these learning objectives refer to the extension of concepts in key stage 2)

Conceptual step	*Example learning objectives*
	Children will:
Revising the basic principles of simple circuits and representing them schematically.	be able to represent their circuits and to construct circuits from diagrams; be able to diagnose if a circuit will work from diagrams;
Learning about electrical conductors and insulators.	know examples of electrical conductors and insulators; be able to test whether or not a substance is an electrical conductor; identify materials used as conductors/insulators and know some of their basic applications; be able to apply this knowledge in their explanation of the construction of circuit components e.g. wires, bulbs and batteries;
A more advanced model for understanding what goes on in simple circuits.	understand that a complete circuit is required to light a bulb or similar and why this is so; know the functions of battery, wires, bulb, switch, motor and buzzer and be able to explain these in appropriate terms; be able to build more complex series circuits to solve technological problems; be able to construct a range of switches, including more complex types such as pressure switches;
Understanding what happens when the number of components in a circuit is changed.	understand a basic model of the relation between the push of the battery, how hard it is for the electricity to flow and the amount of current that flows; know that in a series circuit the current is the same at all points; be able to apply this model in an explanation of what happens when the numbers of batteries/bulbs are varied in series circuits; be able to apply this model in the application of simple variable resistors and fuses.

Conceptual step one – revising the basic principles of simple circuits and representing them schematically

This conceptual step should hopefully be relatively straightforward as it is essentially revision of the learning covered in key stage 1 with the addition of the children being introduced to the circuit symbols and these then being used in the presentation and recording of the various learning activities.

Eliciting ideas

Getting to grips with circuit diagrams is firstly a straightforward matter of learning the necessary symbols and I usually introduce the children to the basic symbols, letting them discuss what they think they might be before giving the correct explanation. After this, it is a matter of letting the children practise building circuits from the diagrams and representing the circuits they design in diagrammatic form. This gives an excellent opportunity to elicit the children's ideas on the basic concepts that they covered in key stage 1 and so, at this stage, I set them circuit predicting, building and explaining tasks to allow pre-assessment of the concepts outlined in the previous section. This is important as, should the children still be confused regarding those concepts, they may well take on board the learning of conceptual steps one and two but only have misconceptions reinforced that will then seriously hamper them when they come to tackle the learning of conceptual steps three and four.

Be warned!

Although you are unlikely to encounter any serious conceptual problems regarding circuit symbols, the children can find it difficult to translate the squiggly jumble of wires on their desk into a neat circuit diagram. This is another good reason to carry out simple revision exercises before getting into more complex content.

Challenging the children's ideas

Since misconceptions are not likely to be a problem, the activities you will need to use tend to be quite straightforward, for example, using the interactive whiteboard to display pictorially represented circuits. The children then have to translate these to circuit diagrams. This can be done on the whiteboard, which can then display the correct diagram for comparison. You can then undertake the exercise in reverse. It is also easy to use such activities for checking the children's ideas on concepts covered previously. For example, you could use the interactive whiteboard to display diagrams of the sort of circuits the children would have constructed in key stage 1 but with some incorrectly constructed (no battery, wires missing or incorrectly connected etc.); the children have to predict if the bulb will light or not and why. They can then build the circuits to check their predictions.

Conceptual step two – learning about electrical conductors and insulators

The learning in this conceptual step is obviously closely linked to the children's understanding of materials and their properties and it is essential that they have a solid grounding in this area. The activities also allow a reinforcing of the models for understanding what happens in simple circuits that the children have been introduced to in key stage 1.

Eliciting ideas

As well as the concepts associated with electricity, it is important to check the children's ideas regarding materials and their properties. Accordingly, I usually begin with discussion of a basic circuit and ask the children to identify materials used in its construction. The discussion can then be moved to the properties of the materials and here the concept of conductivity or, 'letting electricity through', as the children are likely to view it, is likely to arise with little prompting. It is then an easy task to ask the children what materials they think electricity can flow through and why they think this is.

Be warned!

The children often concentrate on metals as being the only electrical conductors and much less frequently suggest other materials such as water, or human tissue.

I used to find that children at this stage would be very hard pushed to suggest reasons why some materials allow electricity to flow easily through them. And, when they did make suggestions, these were often based on misconceptions regarding what electricity actually was. For example one child once told me that, 'materials that let electricity through must have stuff in them like little wires to let the sparks through', whilst another, who was confusing electricity with some sort of fuel-like liquid, told me that, 'those materials [conductors] probably have really small holes inside so the electricity can run through'. I'm glad to say that, since using the techniques outlined in Chapter 16, I find many children are quick to appreciate that conducting materials let electrons move around in them, while insulating materials don't. It is true however, that the children's notion of how an electron moves around is likely to be vague at best and I have had children make statements such as, 'materials that let electricity through must have holes in them so the electrons can get about'. On the other hand, such a statement is not far from the truth and certainly not inappropriate for children at this level.

Challenging the children's ideas

It is easy for children to test their ideas as to which materials conduct electricity by getting them to design a circuit such as the one shown in Figure 17.1.

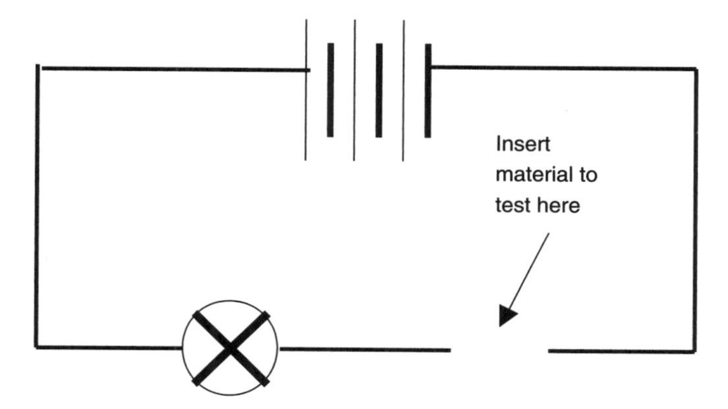

Figure 17.1 An example circuit for testing the electrical conductivity of different materials

Be warned!

Although the children will seldom have any difficulty designing and using such a circuit, there are several things that can go awry with this test, so watch out for the following:

1 Use a 4.5V battery. Less powerful batteries may not have the 'push' to get the electricity through substances that otherwise should conduct electricity.

2 Beware of paint, varnish or layers of oxidization on tins, coins etc. that may prevent conduction and so confuse the children.

3 The children may correctly predict that substances such as their own bodies or tap water will conduct electricity but find that when they test them, they appear not to. This is because the batteries in use in the classroom do not have enough 'push' to make the electricity flow through these materials. More powerful batteries would, but are not safe for classroom use.

Challenging the children's ideas

If you bear in mind the potential problems above, you will find that the children normally cope well with this learning but there are still some issues that you are likely to have to deal with. Even if the children only focus on metals as conductors, it is important that they understand that some non-metals will conduct electricity. A good way of opening their eyes to this fact is to let them predict and experiment with pencil stubs sharpened at both ends. The wood will not conduct electricity but the graphite will.

It is also likely that some of the children will have predicted that human bodies or water should conduct electricity and then have been surprised to find that this was not the case. Water with impurities dissolved in it (in practice any water that hasn't been specially prepared by a process such as distillation) will conduct electricity and this also explains why human tissue conducts electricity; it is largely made up of salty water. However, these materials are not such good conductors as metals and the batteries the children will be using do not have enough 'push' to make the electrons flow through them. It is important not to let the children believe their predictions to be wrong and so I always challenge them to see if they can find other, non-experimental, evidence to back up their original prediction. This they usually easily do and I have had children come up with all manner of evidence from the safety lessons they learnt in key stage 1 to a James Bond film where a villain is electrocuted in a bath.

Conceptual step three – a more advanced model for understanding what goes on in simple circuits

The children should already have had a sound grounding in these areas at key stage 1 but they can now progress their understanding with a more advanced particle theory of electricity applied to how batteries and bulbs work.

Eliciting ideas

Questioning the children during the activities in conceptual steps one and two should provide ample evidence as to the children's understanding of the workings of simple circuits. However, if you have been able to split this work into two units of work, it may have been some time since the children built and discussed circuits and, if this is the case, some quick circuit building and revision will be in order with you questioning the children's ideas on: what is necessary for a circuit to work; what electricity is and how it flows around the circuit; and the functions of the components in the circuit.

> ***Be warned!***
>
> The misconceptions noted for key stage 1 children in the previous section must still be borne in mind, although, if the suggested activities have been followed, these are much less likely to pose a problem. Misconceptions are notoriously difficult to displace however, so don't assume the children have taken on board the correct explanations!

Challenging the children's ideas

The same sorts of basic circuit building activities as were used in conceptual step one can quickly be employed again to revise basic points; to focus in more detail on batteries and bulbs and to look at motors (if the children have not already encountered them).

Motors are easy to introduce and, at this level, there is no need to get into how a motor works but the children should be encouraged to notice that the direction the motor turns changes depending on which way the motor is wired into the circuit. This is easier to see if something like a fan blade is attached to the motor.

It is possible to look at batteries and bulbs in rather more detail than motors and in a manner that will aid the children in understanding exactly what happens in a circuit. In both cases the children will be unable to practically discover the information for themselves but the careful use of models allied to plenty of discussion works very well.

A more sophisticated model of the role of battery

This model depends on the children having advanced their basic particle theory through the sorts of learning experiences outlined in Chapters 8 and 9. They now need to be aware of the fact that materials are comprised of particles called atoms. (An appropriate key stage 2 model of an atom was discussed in Chapter 9.) The most important facets of the model, so far as understanding electricity is concerned, are that: in the middle of the atoms is a fixed nucleus that has a positive charge; whilst hovering around this are negatively charged electrons.

Batteries work as they contain chemicals that, when they react, cause electrons to be pushed around. This is shown in Figure 17.2.

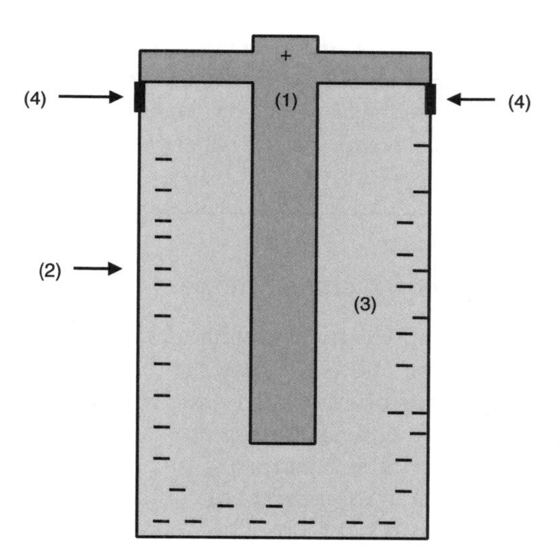

1. The 'lid' and connected 'spike' (1) form one terminal of the battery.
2. The 'pot' (2) forms the other terminal.
3. Special chemicals (3) react with the metals in the battery and push electrons towards the 'pot' meaning it has a surplus of electrons and so is the negative terminal.
4. This leaves a deficit of electrons at the 'lid', making it the positive terminal.
5. The 'lid' and the 'pot' are kept apart by insulation (4) so the electrons cannot flow back to the lid unless they are given an external path through a circuit.

Figure 17.2 A simple model of how a battery works

Be warned!

It might be tempting to cut open a battery so that the children could actually see the structure outlined below (this was exactly the sort of thing I did as a ten-year-old) but the corrosive and poisonous nature of the chemicals involved mean that this cannot be safely done in school.

Like charges repel and unlike attract and this is what pushes the electrons from the negative terminal round to the positive terminal. (The children can make the link to the identical phenomenon they should have encountered in their work on magnetism as was discussed in Chapter 11.)

When the chemical reaction is finished, the battery can no longer push electrons round a circuit.

Be warned!

Rechargeable batteries often cause confusion for the children. Since they are plugged into the mains to recharge them, they tend to reinforce the misconception that batteries are a store of electricity that gets used up in the circuit. It is important that the children understand that when the batteries are being recharged, the electricity from the mains is being used to reverse the chemical reaction in the battery so that it can once more push electrons round circuits.

A more sophisticated model of how electricity flows

It has already been discussed that, in electrical conductors, the electrons can move about relatively easily and this is what allows the material to conduct. Electrons flow through such materials by 'hopping' from nucleus to nucleus. In other materials the structure of the atoms makes it very difficult for the atoms to hop from nucleus to nucleus and these materials are electrical insulators.

It is useful for the children to act out this process of conduction. Place two rows of tables in parallel with a wide space in between. This represents a section of wire. Designate some children as fixed nuclei and space them out between the tables. The other children are designated as electrons. Some can hover around the nuclei and the rest can stand ready to flow through the wire when we imagine it has been connected into a circuit. The electrons can flow through the wire relatively easily with only a minimum of jostling with the nuclei, as in Figure 17.3.

A more sophisticated model of how a bulb works

In key stage 1, the children learnt that there was a pathway for the electricity through the bulb that formed part of the circuit and that the electrons passing along this heated the filament, making it glow and give off light. This basic model can now be extended.

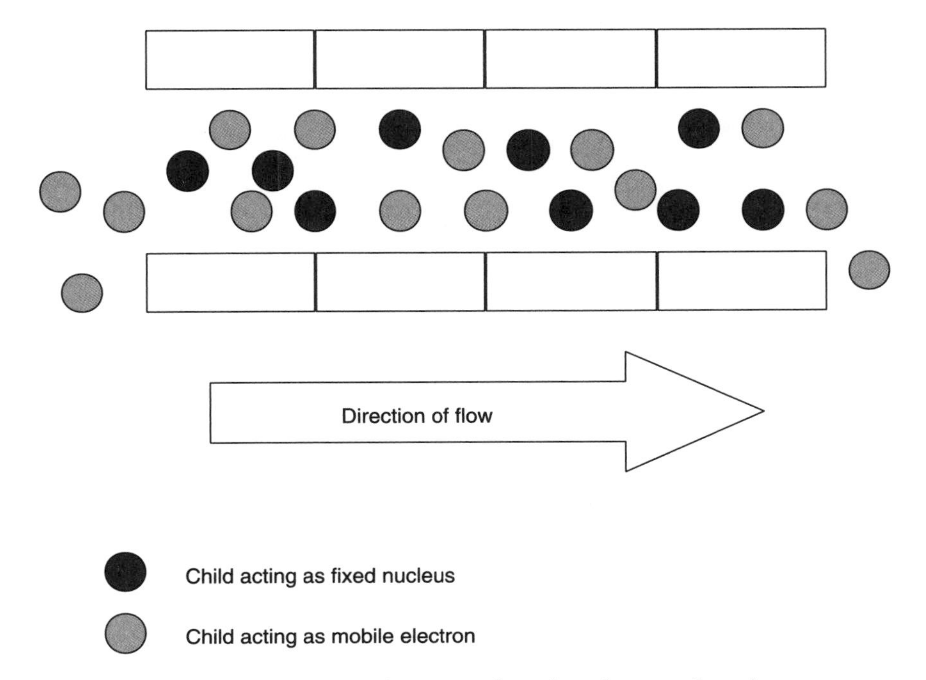

Figure 17.3 Acting out what happens in a normal section of connecting wire

Now that the children have a more sophisticated particle model that includes the presence of fixed nuclei and moving electrons, they can understand how the electrons heat the filament. The easiest way for the children to appreciate this is, once again, to act it out. But first ask the children to make some observations about bulb filaments. The key feature to notice is that they are very thin. And, although it is better to let the children attempt to work out the explanation after acting out the model, this explains why the filament glows: the confined space makes it harder for the electrons to flow and they collide more with the fixed nuclei causing friction. The friction generates heat and the filament glows white hot and gives off light.

The children may also observe that many filaments are wavy. This increases their length and therefore the number of collisions between electrons and nuclei, so giving more light. They should also be encouraged to speculate about the material the filament is made of and to realize it must have a high melting point. In fact most filaments are made of tungsten for exactly this reason.

Now model what happens in the filament. This is done in a similar way to how the children modelled the flow of electrons through a wire as in Figure 17.3. However, the filament is much narrower than an ordinary piece of wire and so the tables must be pushed closer together, squeezing up the fixed nuclei. Figure 17.4 shows that now it is harder for the electrons to flow and there are many collisions with the nuclei. If you ask the children what happens when objects collide against each other, they will hopefully make the link to friction and how this generates heat. (Getting them to clap their hands slowly and then quickly, may help get this concept across.) It is the heat generated from these collisions that causes the filament to glow and give off light.

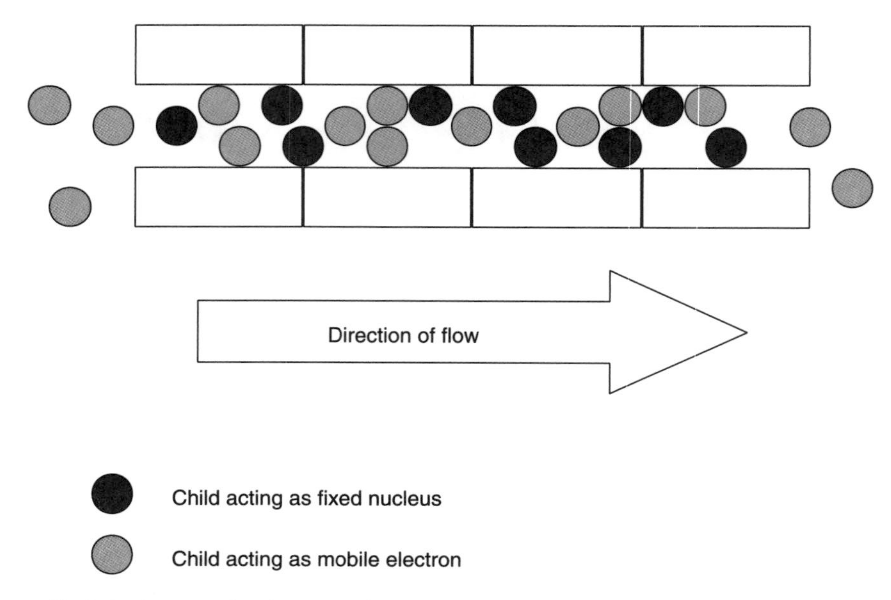

Figure 17.4 Acting out what happens in a narrow bulb filament

The model also introduces the concept that it is harder for the electricity to flow through some components than others and this will be important when the children come to vary the number of components in a circuit and observe the effects.

Be warned!

No model is perfect, otherwise it would be the actual thing it is trying to represent! The major problem with this model is that it is likely to give the impression that the flow of electrons slows down when it is harder for them to flow. This misconception arises from the fact that the children naturally slow down when they are trying to squeeze through the narrower gaps. In fact, it is the opposite that happens in filaments but this utterly counter-intuitive fact can be left until the next conceptual step.

Hint!

One tactic I have tried to counter the problem of the 'electrons' slowing down when using the model is to explain to the children that we are going to act out what happens in very, very slow motion. The electrons are then only allowed to take a step every time I clap. This helps ensure a constant speed but the children still readily appreciate that there are more collisions in the narrow filament.

The model can also cause problems as the 'electrons' are likely to bunch up when flowing past the fixed nuclei, whilst actual electrons will be spread evenly through the wire. To avoid this becoming too obvious, I use a relatively short length of tables and ensure that the children are fairly densely placed along its length. This keeps them crowded enough so that bunching is less obvious. The 'take a step when you hear a clap' method will also help prevent this problem.

Be warned!

You should also encourage the children to think about what is inside the bulb. Their work on materials should let them appreciate that it can't be oxygen otherwise the filament would burn, but be careful! It is a common misconception that there is a vacuum inside the bulbs. In fact they are usually filled with argon, a gas which does not react with the hot filament and helps make the bulb stronger.

Conceptual step four – understanding what happens when the number of components in a circuit is changed

This area represents the culmination of primary school children's work on electricity and it often causes severe problems. In fact, I have seen no other area of primary science result in such conceptual problems. This need not be cause for panic, however, as the problems stem not from the children's inability to grasp the concepts but from our being unsure how to present the learning activities. The outline below will help you avoid the many hazards.

Be warned!

If you are not exactly sure what you are doing, it is very easy to go wrong with these activities and so I have outlined them in some detail. This may seem like a rather programmed approach but this is the best example of where such tactics are required in this topic. The learning can still be presented in a way that allows the children to connect up with their own ideas but, if you give a free hand to the children to explore this area, it will almost certainly result in reinforcing misconceptions and the children discovering phenomena that are extraordinarily difficult to explain in terms accessible to the children.

Eliciting ideas

Show the children the circuit diagram in Figure 17.5 and ask them to predict how the bulb will appear.

The bulb will be very bright, indeed it is a good idea not to leave the circuit connected too long as the bulb may blow (see the warnings regarding bulbs at the end of this chapter).

Now ask the children to predict how the bulbs will appear in the circuit in Figure 17.6 and to explain their reasoning.

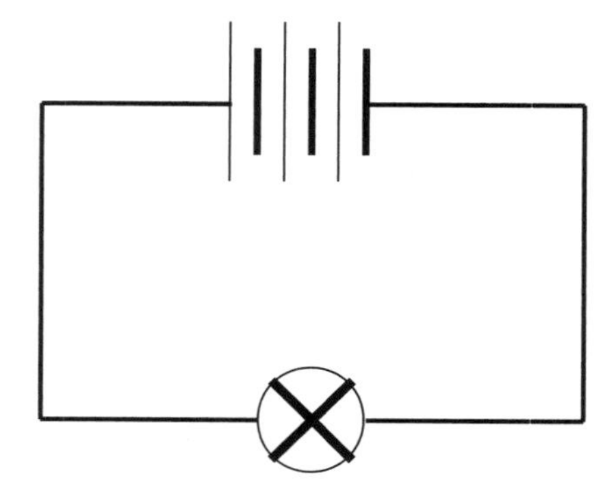

Figure 17.5 A circuit with 4.5V battery and one 2.5V bulb

Be warned!

It is imperative that you use identical bulbs for this activity. If you use bulbs of different voltages they will not be of the same brightness and explaining why is beyond children at this level. Even amongst 2.5V bulbs, there can be manufacturing differences such as a different colour to the base of the bulb that will lead to the bulbs appearing to be different in brightness.

Hint!

I find the safest option is to use fresh bulbs from the same packet, and when doing this, I have never had any problems. The bulbs can then be re-used in follow-up technology activities.

The two bulbs in the circuit in Figure 17.6 will be dimmer than the bulb in the circuit in Figure 17.5 but will be of equal brightness compared to each other.

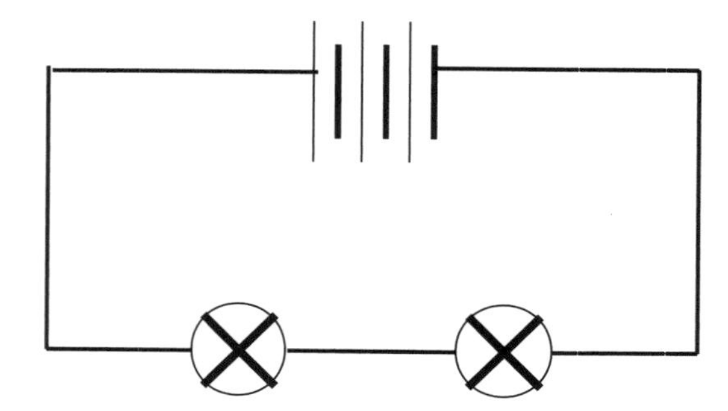

Figure 17.6 A circuit with 4.5V battery and two 2.5V bulbs

Be warned!

I used to find a consistent pattern of predictions from the children. Typically, they correctly predicted that the bulbs in the second circuit would be dimmer but often gave incorrect reasons such as, 'the electricity has to be shared between the two bulbs'. They then frequently showed they believed that the two bulbs would be different in brightness and made statements such as, 'the first bulb uses up some of the electricity and so the second one isn't so bright'. I am pleased however that, since I began using the careful sequence of activities outlined here and in Chapter 16, I have found that these misconceptions rarely appear as the children appreciate that the electrons do not get used up and that the same number flow around all parts of the circuit. Now I find the children correctly predict that the two bulbs will be of equal brightness but that they become confused as to the difference in brightness between the first and second circuits. They instinctively feel those in the second circuit should be dimmer but they are often at a loss to explain why as they have not yet encountered a model to explain why the flow of electricity should decrease.

Challenging the children's ideas

The next stage is to allow the children to test their circuits and build a model to explain their observations.

The children must then explain why the bulbs are dimmer in the second circuit but the two bulbs are of the same brightness. The second factor should not present problems, as the earlier activities will have shown that there is the same number of electrons going through both bulbs. The first part is trickier. The children can be reminded that bulbs make it harder for electricity to flow and I find that this leads to their suggesting two possible conclusions: the electrons in the second circuit are moving more slowly; or there are less electrons moving in the second circuit. There is no practical way to check this at primary school (even an ammeter could be interpreted in terms of either explanation); the best solution at this stage is to refine the model for what happens in a circuit.

A refined model of electricity in a series circuit

This is similar to the previous model first introduced in Chapter 16 but incorporates a refinement to allow the children to appreciate how the flow of electrons can vary between circuits.

Step one

Scatter Multilink cubes in a large ring around a table. These are to represent the electrons present in every part of the circuit. It's best to have lots of cubes and scatter them roughly evenly. Each cube will actually represent a enormous number of electrons.

The children then form a ring around the table. They are to represent a loop of wire with one child designated as a bulb and given an appropriate sign to hang around their neck. This child can be given a crown to indicate when they are lit.

Will the bulb light? No, there is no battery and so no electrons are pushed around the circuit.

Step two

One child is designated as a battery and given an appropriate sign.

Will the bulb light now? Yes, as electrons are pushed around the circuit from the negative terminal of the battery to the positive one.

BUT the battery does not have enough 'push' to move every single electron in the circuit. Simulate this by having the children lift three cubes each and pass them hand-to-hand in the appropriate direction.

Hint!

This bit can get a little chaotic and so, to help the children pass the cubes evenly, so that at any given time each child has three cubes, I find it best to clap slowly with the children passing the cubes on each clap. It also helps avoid the scattering of rogue electrons if the children link their three cubes together.

'Disconnect' your circuit.

Thus far the model revises three key points:

1 the electrons flow from negative to positive;
2 the same number of electrons are flowing at every point in the circuit;
3 and the electrons do not get used up or destroyed.

It also introduces one new concept: not all the electrons in the circuit are pushed round by the battery.

Step three

Explain that another, identical bulb is to be added to the circuit and ask the children what effect they think this will have. From their experiment they will know that the two bulbs will be dimmer than the one but that they will be of equal brightness. They should also be able to speculate that the bulbs make it harder for the electricity to flow because of the very narrow filaments.

Two bulbs make it harder for the electricity to flow than one, therefore the battery can push fewer electrons around the circuit. This time allow each child to lift and pass just two cubes.

The model now explains the results of the children's observations:

1 less electrons are flowing so the bulbs are dimmer;
2 but the same number of electrons are flowing through each bulb so they are equally dim.

Step four

You can now get the children to experiment with adding further bulbs which will mean less electrons will flow, or, if too many bulbs are added there may not be enough push to light any bulbs at all. They can also add extra batteries which will mean more electrons will flow making the bulbs brighter, but the bulbs will always be equally bright, as the same number of electrons flow through each.

Be warned!

As is the case with any model, there are problems. It is, of course, hopelessly out of scale. Nor is it intended to realistically represent the actual ratios of changes in flow of electrons. However, I have found this model to be the most successful of any I have used to attempt to help children learn these concepts.

Consolidating the lessons of the model

The concepts involved here are complex and it is useful to spend time ensuring the children consolidate their learning.

What the children have now discovered is that there is a relationship between the 'push' of the battery (measured in volts), how hard it is for the electricity to flow (technically this is called the resistance of the circuit) and how much electricity actually flows (the current). Physicists would measure the resistance of the circuit in ohms and the flow of electrons in amps but there is no need for the children to actually measure such quantities in primary school and therefore no need for the terms to be introduced. I do use the term volts with the children as this has a practical significance for them. They will see it marked on batteries and readily accept it as a measure of the strength of the battery's 'push'. They will also accept the voltages stamped on appliances etc. as a guide to how much battery push is ideally required to operate them.

In effect, the children have now discovered a basic form of what in physics is know as Ohm's Law. Learning the details of Ohm's Law is most definitely out of place in the primary school but the level of understanding illustrated here is appropriate. I have found problems where curricula sometimes state that resistance need not be taught. Often the same curricula specifically state that children should vary the number of components such as bulbs in a series circuit. If the children are to do this, they cannot possibly explain their observations without reference to resistance, which reduces the activity to dangerous guesswork that will leave misconceptions unchallenged. The concept of resistance is not a difficult one and I have never found the children to struggle with it when a careful programme of activities such as those outlined are followed. You can help the children consolidate their learning by asking them to build and compare circuits as outlined in Figure 17.7.

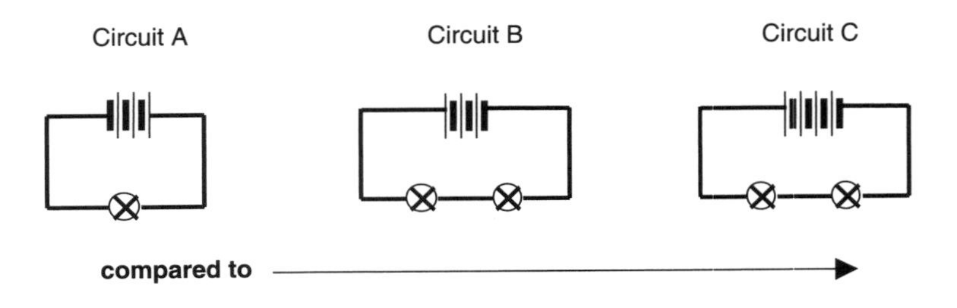

Circuit A	Circuit B	Circuit C
Relatively large push	Same push	More push
Relatively little resistance	More resistance	Same resistance
Therefore large flow	Therefore less flow	Therefore more flow
Therefore bright bulb	Therefore bulbs dimmer but equal	Therefore bulbs brighter and equal

Figure 17.7 The relationship between push, resistance and flow

Be warned!

The children often think that the electrons slow down when they reach the bulbs as the filaments are harder to flow through. Unfortunately this misconception is often reinforced by many models often used to help children understand electricity. (A warning about this was given when discussing the model introduced to explain why a bulb lights, see Figure 17.4.) Thinking carefully, and bearing in mind that there are the same number of electrons flowing in any part of the circuit, will show that the electrons must not be slowing down, otherwise they would bunch up in bottlenecks when going through bulbs etc. but this is a level of abstract thinking that would seriously challenge many children at this level. The answer lies in the fact that adding a bulb doesn't just make it harder for the electricity to flow through the bulb; it makes it harder for the electricity to flow through the whole circuit. I have found that this is a concept that the children can more readily grasp. You can illustrate this by getting the children to form a circuit and grip a hoop to represent the electrons which flow in a continuous chain around the circuit. If one child is designated as a bulb and grips the hoop more tightly, all the children will find it more difficult to move the hoop. The problem with this model is that it implies that increasing the resistance slows the flow of electrons, rather than reducing the numbers flowing, so I only introduce it after the children are thoroughly familiar with the model that uses the Multilink cubes introduced above and so understand that it is the number of electrons flowing that varies, not their speed.

Further notes on varying the numbers of components in circuits

When undertaking this sort of activity, it is best to stick to varying only identical bulbs or the 'push' of the batteries. The effects of these variations are easy for the children to explain: more bulbs mean it is harder for the electrons to flow, therefore less electrons flow; more batteries mean more push, therefore more electrons flow.

The children may suggest varying the length of the wires in the circuit and this is possible. Pay careful attention to their predictions as to what will happen. I have had children suggest to me that the extra length of wire will 'slow down the electrons, so the bulbs won't be so bright', thus showing they have not fully grasped the concepts outlined above. On the other hand children have made predictions such as, 'it's like the long bits in the bulbs [the filament] the longer wire will make it harder for the electrons to flow so there'll be less and the bulbs will be dimmer'. This prediction is correct though the children may need to use very long lengths of wire indeed before they notice a difference in bulb brightness. To extend capable children I have also sometimes allowed them to vary the type of wire used. The best variation is to use high resistance wire, which is easily available from educational suppliers. I have found that the children readily notice that it is thinner and correctly link this to higher resistance, as in light bulb filaments. They also often, again correctly, speculate that it is made of different metal and that this might make it harder for the electrons to flow.

What you really want to avoid is allowing the children to vary components e.g. combining bulbs with buzzers. If they do so they will get some weird results, the explanation for which is really beyond the children and which may lead to misconceptions. As a safety net and example of just how powerful the models and explanations outlined in this section are, it is useful to consider just what happens in such nightmare scenarios.

I was once horrified to observe a student set some children the task of predicting what would happen in the circuit illustrated in Figure 17.8 and then ask them to build the circuit to test their predictions.

The children had already built the circuit with just a bulb in it but had not had the benefits of the sort of models to help their understanding as outlined in the preceding

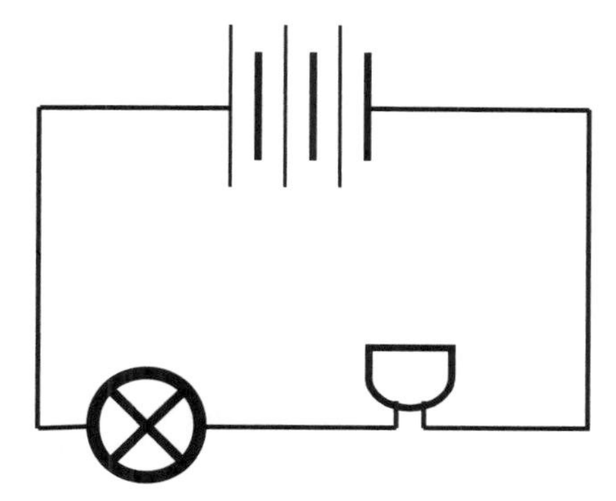

Figure 17.8 A circuit with 4.5V battery, 3V buzzer and 2.5V bulb

sections. Accordingly it was no surprise that they uniformly gave evidence of various versions of the misconception that electricity would be used up in one appliance and so one would work better than the other. For example, one very sharp child who had, correctly, heard that electricity came from the negative end of the battery predicted, 'I think the buzzer will be loud because it's near the negative bit but the bulb won't be bright because the buzzer will use up a lot of electricity'. This is excellent reasoning but, of course, scientifically incorrect! A terrible problem then arose when the children built the circuit and discovered that the buzzer buzzed and the bulb didn't light at all, thus seemingly confirming their predictions and thus their misconceptions. How could the children's misconception be challenged and the correct explanation given? The activities outlined above were effective at challenging the ideas but there was a great deal of work involved. As concerns the scientific explanation of why the bulb didn't light, it is closer to A-level standard than primary school work, but I was extremely pleased that the models we subsequently introduced the children to allowed them to advance a simple and correct explanation that satisfied them. They concluded that, 'The buzzer and the bulb together made it harder for the battery to push the electrons; therefore less electrons went around the circuit. The number of electrons going round was enough to shake the buzzer and make a noise but not enough to heat up the filament to make it glow'.

Further challenging and extending the children's ideas

It is useful to look at two other fun activities and see how they can consolidate the models introduced.

The first activity is very simple. The children have already learnt that graphite is a conductor and, if they introduce a split pencil into a circuit containing a 4.5V battery and a 2.5V bulb they can vary the position of the crocodile clips connecting onto the pencil. This will vary the resistance, and thus the flow of electricity, and thus the brightness of the bulb.

The second activity is also simple but involves a piece of wire wool momentarily getting hot and melting. There is no serious risk in this but it is better to demonstrate it to the children rather than have them carry it out themselves.

Ask the children to consider the example shown in Figure 17.9. Is there electricity flowing?

The answer of course is yes: there is a complete circuit plus a battery so electrons will be flowing. What is more, since there is a relatively large 'push' and relatively little resistance, there will be a relatively large flow of electrons (current). This will accelerate the chemical reaction and quickly leave your battery useless! All those electrons colliding with fixed nuclei in the wires will also heat the circuit up, although not dangerously so in this case. Such short-circuits can be dangerous however and this can be demonstrated by placing a strand of wire wool between the crocodile clips as part of the circuit. A fresh 4.5V battery will have enough push to generate a current large enough to quickly melt the wire. This looks spectacular but is quite safe if carried out as a carefully supervised demonstration. The children can then apply this to some research as to how fuses work and why they are such an important safety device. You can also encourage the children to link their observation back to their

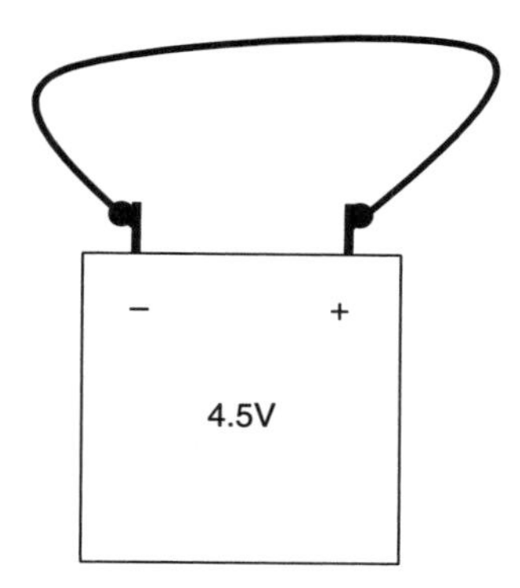

Figure 17.9 A 4.5V battery short circuited

understanding of light bulbs and why they cannot be filled with air as this would contain oxygen that would react and burn with the hot filament. They will notice that the wire wool does not just melt but burns with a brief, tiny flame.

Choosing equipment to help children learn about electricity

Electricity is really the only primary school topic where much specialized equipment is required. Choosing this equipment is vital as poor choices can easily lead to creating conceptual problems for the children. It is also an area that often does not get the attention it deserves in the various support materials designed to help teachers. This section will help you avoid the mistakes that I used to make and that I have found many teachers and students making since.

Choosing the equipment

If your school is one where the electricity equipment is stored in a jumble in an old cardboard box, you are in trouble! Selecting the right kit and storing and organizing it properly are vital. The first step is to consider what equipment to use.

Batteries

Dry batteries are the safest type for use in primary schools. Wet batteries such as car batteries or accumulators should not be used as they are too hazardous.

For key stage 1 work 1.5V, D size is the easiest to handle. (N.B. these are technically *cells*, rather than batteries, but in primary school this distinction is not important.)

For key stage work, the 4.5V, 3R12 size is the best bet as many of the circuits the children will be building will require extra ‘push’ from a bigger battery. (N.B. see the warning under the bulb section about disconnecting circuits when not in use.)

Fresh batteries are important as if they have insufficient ‘push’, the children may encounter results that confuse them and lead to misconceptions. I always have fresh batteries for each topic but then re-use them for D&T projects that the children can take home.

Be warned!

Rechargeable batteries are advised against, as they can lead to overheating if discharged very quickly, for example in a short circuit.

Battery holders

Suitable holders will be needed for the 1.5V batteries to allow the children to connect them into circuits easily.

Bulbs

Choosing the correct bulbs is crucial. I advise using 2.5V bulbs for all primary school science. These can be lit by the 1.5V batteries and can be used in multiples with the 4.5V batteries at key stage 2.

Be warned!

Care should be taken when there is just one bulb in a circuit with a 4.5V battery, as it may 'blow' quite quickly. It is a good idea to get the children into the habit of disconnecting their circuits as soon as they have made the necessary observations. This will also prolong battery life.

Bulb holders

These will be needed to allow the easy connection of bulbs into circuits.

Wires

Buying commercially produced lengths of wire ready-attached to crocodile clips is worth every penny. Even those children who can cope with cutting, stripping and connecting loose wires take ages to do so and often end up with loose connections that can cause conceptual havoc.

Be warned!

The insulation on such wires normally comes in a variety of colours. Many children believe the colour of the insulation on the wire is significant and you should challenge this misconception by letting the children swap around the different colours to see if there is an effect. Alternatively you might sort the wires so that each group of children receives wires of only one colour.

Switches

As was discussed, the best way for the children to learn about switches is for them to build their own. Having some commercially produced switches can be useful, however.

> **Hint!**
>
> Choose switches where the children can actually see how the switch works by breaking and connecting the circuit. The 'doorbell' type is usually quite good for this as the children can see the metal tabs behind the fascia of the switch.

Buzzers

Yes, they are a pain! But they are fun for the children and useful for showing electricity doing other things than lighting bulbs. Choose the 3V ones, your batteries may not have enough 'push' to buzz the 6V sort.

> **Be warned!**
>
> Buzzers will only work if correctly connected (the black wire goes to the negative terminal of the battery). Make sure the children discover this – I once had to stop a student throwing out buzzers that the children claimed 'didn't work'.

Motors

Choose motors no larger than 3.5V, or, once again, your batteries may not be able to work them.

> **Hint!**
>
> In all cases, especially batteries and bulbs, it is advisable to have to hand an appropriate stock of spares. Sometimes components do break and you will not want this to hold up the children's learning.

Storing and organizing equipment

This is another area where inexperience or carelessness on our part can impede the children's learning. I have found the following principles very helpful.

Don't – store the equipment in tool boxes with the idea that a group of children can take a box and select from it what they need for a given activity. The children will be presented with more equipment than they need and will become distracted with potentially serious conceptual consequences.

Do – carefully store the equipment centrally in a manner that clearly separates the various components and that can easily be checked to ensure all is in the correct place.

Do – present the children only with the equipment they need for any given activity. For less experienced children I usually give them the equipment in a small tray. More experienced children can take a tray and collect the appropriate equipment.

Do – provide the children with pictorial or symbolic checklists of what equipment they are using. This allows the children to check they have the correct equipment when setting up and tidying away and helps to familiarize older children with the correct circuit symbols.

Don't – use commercially produced electricity kits. These are an expensive and inflexible approach to the topic. The guidance materials that come with them rarely focus on the children's ideas. Worst of all are the sets that have a circuit board into which various components can be plugged. Such kits obscure the workings of the circuit, which hinders the children's understanding.

Do – (and here we have an exception to the above) consider the kits that are specially produced to support children with manipulative difficulties. These have the normal components set in easy to handle and connect mounts and can be a great help for children whose dexterity is poor. It is important however to employ the kits in the same carefully structured manner that you apply to normal equipment.

Do – specifically teach the children self-help skills to employ if they can't get their equipment to work. For example get them to check they have connected everything properly before they come tugging at your sleeve claiming their bulb doesn't work! BUT, remember, you can only really start this process once you have pre-assessed the children's ideas regarding what is required for a circuit to work and they have tested these out.

18

Developing an initial understanding of forces

Introduction

Research shows that forces is one of the most difficult primary school science topics. Many of the concepts are counter-intuitive and there are a host of common misconceptions that can be extremely difficult to challenge. These difficulties beset not just the children but often ourselves as teachers. Even when we are confident with our own subject knowledge in the topic, it can be very difficult to know how to make the concepts accessible to children. However, I have found that by taking care over when and how concepts are introduced and by following a few simple rules to guide thinking about forces, many of the pitfalls can be avoided.

Where this topic fits in

There is a very large spread of conceptual progress in this topic and I would not attempt to cover it in less than three stages in primary school. One slot at key stage 1 is sufficient. At this stage, with the exception of a few explanatory concepts, the learning is relatively straightforward and so the topic can be addressed either early in the key stage, or later. A lower key stage 2 slot is then useful for extending understanding and this slot can also be used to cover the more complex key stage 1 concepts, if the topic was first introduced to younger children. Finally an upper key stage 2 slot is needed to deal with the most difficult concepts.

It is worth noting that I have found that there is potential for considerable overlap between some of the mid-range concepts in terms of their being introduced to key stage 1, or key stage 2 children. Many curricula and schemes suggest that work such as the first stages of measuring forces, be introduced in lower key stage 2 but I have found that these concepts are perfectly accessible to upper key stage 1 children, so long as they have a strong learning context that helps them make sense of their learning. Accordingly, if your school is covering key stage 1 forces in upper key stage 1, it might be appropriate for your children to engage with the learning outlined for the first two conceptual steps introduced in Chapter 19.

Links to other topics

If you have read this far in the book, you will not be surprised to learn that a good grounding in materials and their properties is essential for the children to explain some of the concepts they encounter. Accordingly, before commencing work on forces, it is useful if key stage 1 children have covered the concepts outlined in Chapters 6 and 7; and key stage 2 children have covered the concepts in Chapters 8 and 9.

A context for the topic

Toys are often suggested as a suitable topic for introducing young children to forces and indeed this is true; they are fun; familiar to the children; and present the concepts in an accessible way. Likewise, I have found that a PE based topic based around teaching the children new physical games can work very well. It is possible to go a bit further than this though and link the learning of forces to some really great cross-curricular work.

Once, when working with some year 1 children on the topic of light, I had organized an 'underground caving expedition' for them that involved us dressing up and exploring the gloomy depths underneath the stage in the school hall. The next year, I was doing some more science with them, this time on forces, and the children insisted they wanted to be explorers again. It is good practice to allow the children to base their learning on their own ideas and so the learning context for the work on forces became 'explorers'. This worked extremely well. The major cross-curricular elements were lots of work on language, as we engaged with, and created our own, explorer narratives; plus some excellent geography, as we learnt about far away places and cultures. From a science point of view, there was scope for, not just introducing the concepts, but also for excellent investigative work, as we had to solve the various problems of transporting ourselves and our equipment through all kinds of exciting locations from the jungles of equatorial Africa to the icy wilderness of Antarctica. The value of a good learning context was demonstrated by the fact that without any artificial forcing on my part, and without the need to allocate substantial extra time to the topic, the children began to successfully engage with forces concepts such as quantifying forces, that are often considered as better introduced in lower key stage 2.

Conceptual step one – observing that things speed up, slow down, change direction and change shape

The core concept that primary school children require to get to grips with forces is an understanding of what forces can actually do, as it is this that allows them to spot where forces are acting. This means that this first conceptual step, even though it appears very basic, is extremely important.

Eliciting ideas

In many ways, this first conceptual step is an exercise in observation and vocabulary and it is important to check how well the children can observe and describe the

Table 18.1 Conceptual steps and example learning objectives for developing an initial understanding of forces (all of these learning objectives are suitable for coverage at key stage 1)

Conceptual step	*Example learning objectives*
	Children will learn:
Observing that things speed up, slow down, change direction and change shape.	to describe objects starting to move and speeding up; to describe moving objects slowing down and stopping; to describe moving objects changing direction; to describe objects changing shape;
Developing a model of forces based on what they do in everyday situations.	to differentiate between pushing and pulling; to correctly identify pushes and pulls in everyday situations; that non-living entities can exert pushes and pulls; that some pushes and pulls do not involve direct contact; to identify the pushes and pulls when objects speed up, slow down, change direction, or change shape and know that pushes and pulls are forces; that the greater the push/pull, the bigger the effect; that a heavier (technically more massive) object will require a greater push/pull to achieve the same effect compared to a lighter object; that forces can be acting even when an object is at rest.

movement of everyday objects. To do this, I get the children to discuss all the possible ways they might travel as explorers. They come up with a most imaginative list including everything from walking to space rockets. We then gather together, or make, models of the various means of travel and make an interactive play display of them. One of the uses this is put to is to have the children describe how the various means of transport could move. This gives me an excellent insight into the children's understanding and description of movement.

Be warned!

Obviously the children's vocabulary will vary but you are unlikely to encounter conceptual difficulties regarding the children's ability to simply describe movement. However, there is one potential misconception lurking behind this activity. Very deliberately, I have been emphasizing moving objects to build up the children's vocabulary. Care must be taken with this approach as, if left at this, it may reinforce the very common misconception that when objects are at rest, there are no forces acting on them. However, this can be a tricky concept for young children and, at this stage, I am content to leave this concept but to be aware that many are likely to have this misconception and that it must be challenged after the initial ground work is completed.

Challenging the children's ideas

The first aim through such activities is to encourage the children to notice the many different ways in which objects move and to ensure they have adequate vocabulary to describe these movements. Through discussion, it is then important to help the children focus on the fact that all these movements can be put under one of four headings:

1. Objects starting to move/already moving objects speeding up. (Technically this is acceleration but at this stage, 'getting faster' is fine.)
2. Moving objects slowing down/stopping moving. (Technically this is deceleration but at this stage, 'slowing down' is fine.)
3. Moving objects changing the direction of their movement.
4. Objects changing their shape.

Be warned!

This seems straightforward enough but there are two problems you will have to watch for.

Firstly, the above taxonomy is very generalized and rather abstract. Very few children of this age will reach this level of generalization without substantial scaffolding. Often you will have to give them the four categories and then give them varying levels of support in assigning the different movements they encounter to the correct category.

Secondly, and allied to the above, the children will often think that a stationary object starting to move is fundamentally different from an already moving object further speeding up; and, in the same fashion, will believe stopping to be different from simply slowing the speed of movement. It is important to try to help them see that both of the former are examples of speeding up, whilst both of the latter are examples of slowing down. However, if some children can't appreciate this and end up with a six category taxonomy at this point, it is not a disaster but it should be noted that they will need further help with this, when they next encounter the topic.

Hint!

It is a good idea to ensure changing shape is sufficiently emphasized, as I have found that this concept can be easily missed. Some contexts can make this harder than others. The example context of means of travel for explorers fitted well with the children's interests but didn't immediately seem fruitful for introducing changes of shape. However, with a little thought, even this context allowed such work. For example we noticed that toy trains changed shape as they went from straight stretches of track to curved stretches. Some of the children's more original travel suggestions also helped. For example one boy had suggested that explorers could bounce about on 'space hoppers'! This was hardly a very practical mode of transport but it nicely facilitated discussions on changing shape as the hoppers squashed and expanded whilst bouncing.

Conceptual step two – developing a model of forces based on what they do in everyday situations

This step involves the bulk of the work at key stage 1 and involves tying together a number of strands into a model that will allow the children to understand how forces influence the world around them.

The first step is to help the children appreciate that all the various actions resulting in movement that they have been discussing can be divided into just two classes: pushes and pulls.

Eliciting ideas

There is no great conceptual depth here to probe the children's ideas on but I normally set them the challenge of seeing if they can combine all the different verbs for the actions that result in movement into as few headings as possible. There is good language work here: first the children have to think of all the verbs that result in the different movements they have been looking at; then the discussion about how to use fewer verbs usually raises interesting ideas. Even so, I usually find that considerable scaffolding is required if the children are to appreciate that all of these actions are either a push, or a pull.

Be warned!

Even if there are no entrenched misconceptions here, there are a number of problems that can confuse the children.

1. Some sources suggest that forces should be classified as pushes, pulls, or twists. The finer points of this would be interesting to discuss with older children but, in practice, I find that this division only confuses younger children and is conceptually unnecessary. Any 'twists' the children do encounter, for example twisting a rope of play putty between their two hands can easily be explained in terms of pushes and pulls; the 'twist' is the result of one hand pushing the putty in one direction, whilst the other hand pulls the putty in the opposite direction.
2. It is also necessary to take care that the children can translate the various actions into pushes and pulls. It can take a lot of practice and some children even find it hard at first to differentiate between what is a push and what is a pull.
3. Watch out for the common belief that only living things can push, or pull things.
4. Another misconception that I have found quite common amongst younger children is that there has to be contact between solids for there to be a push, or pull.

Challenging the children's ideas

The main thing here is plenty of practice to help the children differentiate between pushing and pulling and to correctly identify pushes and pulls in everyday situations. I find that it is a good idea to let the children approach this learning through several different activities. It is also vital for the children to specify what the push, or pull, does, that is: speed up an object; slow down an object; change the direction of movement of an object; or change an object's shape. For example, in the exploring context, the children first identified all the pushes and pulls involved in moving the models of different sorts of transport and what they did. Then, in music, we learnt sea shanties which we sang whilst pretending to load our expedition ship with all the equipment the children suggested they might need. As we loaded the ship, we looked for what pushes and pulls we used and what they were doing, then made up our own shanty chants to suit the actions. This activity shows what a bit of cross-curricular imagination can do, as it is one of the most successful activities for learning about pushes and pulls that I have ever used (we even built a pretend capstan out of broom handles and an old plastic storage bin).

Hint!

Assuming the children's language skills are up to it, a useful exercise I have found to help the children in this area is to get them to write sentences describing their actions. They are then given a selection of stickers with push and pull written on them; and a selection of stickers detailing what the pushes and pulls can do. The children then have to stick the appropriate stickers under the relevant bits of their sentences, as in Figure 18.1.

It is important to let the children explore examples of non-living materials exerting pushes and pulls, being sure that the materials involved include liquids and gases, as well as solids. This is easy in the example context, as the children's imagination

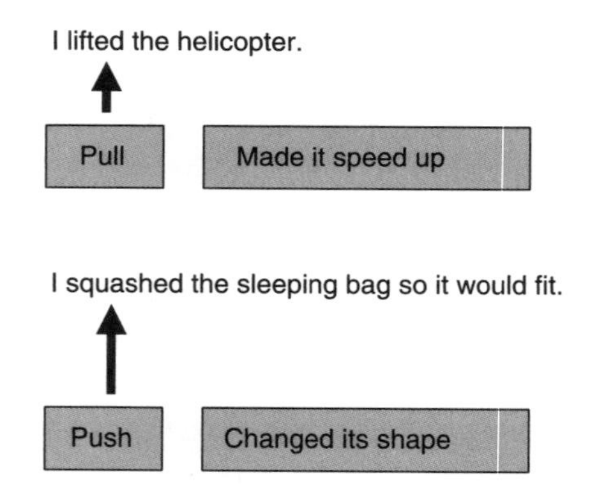

Figure 18.1 Example sentences showing how forces stickers can help children identify pushes and pulls and what they do

has been stimulated by many of the tales of water-borne exploration. Activities such as building model canoes from natural materials and sailing them on the local stream, and investigating if the size of the sail makes a difference to how fast a model boat sails allow the children to appreciate exactly these sorts of concepts.

Simple investigative work (see Figure 18.2) also allows the children to explore the effects of varying the strength of pushes and pulls and to appreciate that heavier objects require greater pushes and pulls to move them. (Technically, one should say more massive objects but the concept of mass is beyond children at this stage.)

By this stage the children should understand what I sometimes term the basic forces facts:

1 That forces are either pushes, or pulls;
2 That forces can only do four things: speed objects up; slow objects down; change the direction of an object's movement; change an object's shape;
3 That the four things listed under point 2 cannot happen without at least one push, or pull to cause them.

This is very useful, not just as it stands, but as an aid to detecting where there are forces at work, even if they are not obvious. For example, most children will readily appreciate that you need a push to throw a ball in the air but it is not surprising that, given the invisible, non-contact nature of gravity's pull, they often think that a force is

Figure 18.2 Investigating the effect of greater pulls on canoes with loads of various weights

not needed to make the ball fall down again. Gravity and falling are complex concepts and I do not aim to tackle them in depth with younger children but it is useful to begin to challenge their ideas. Falling is an example where it is possible to do this. When a ball falls, it is easy for the children to appreciate that it has changed direction (it has actually also slowed to a momentary stop and then speeded up again, but it is the direction change that young children will most appreciate). Using the logic of our forces facts: if the ball has changed direction, there must have been a push, or a pull that has done this. The ball can't 'just fall'. I have found that, if children have had plenty of practical opportunities to appreciate the forces facts, they are much quicker to appreciate this and that it is possible to introduce gravity successfully at key stage 1, though only as an amazing pull from really, really big (technically massive) objects such as planets, stars and moons. Such grounding will help avoid conceptual traps later. Gravity is also an excellent way of introducing non-contact forces. Magnets are also good for this, as discussed in Chapter 11.

The final concept that is worth adding to the children's model is that forces can also be acting, even when an object is not moving. Once again, this is complex for younger children but I have found that it is at least possible to successfully introduce the basic concept, even if the children do not fully understand how exactly the forces are balancing. If you have discussed gravity as above, this makes a good starting point. For example, in the exploration context, the children had played a lot at such activities as loading boats, sledges etc. Here the emphasis was on identifying pushes and pulls and their effects, but it also led to much discussion about the concept of weight, or 'heaviness', as the children often called it. The children quickly appreciated that when they held a relatively heavy object (with due attention to health and safety issues), it pushed down against their hands, even when it wasn't moving. In key stage 1, many children won't progress beyond this but at least there has been something to begin to challenge the belief that there are no forces acting if an object is at rest. However, I have noticed that, even at this stage, some able children will appreciate that the push against their hand is due to the fact that the object is being pulled by gravity, which is what gives it weight and they are then able to reason that, if the object is set on a table, it will also push the table. Some, if they have had a good grounding in materials and simple particle theory, are even able to understand how the table can push back, but this is really more of a key stage 2 concept and is discussed in the next chapter.

Be warned!

A common activity that is often suggested at this stage of the children's learning is to have the children explore different objects and materials regarding how their shape can be changed. It is usually suggested that the children classify the objects into classes such as: objects they cannot change the shape of just using their hands e.g. a pebble; objects that they can change the shape of and that stay in their new shape e.g. a torn, or folded piece of paper; and objects that they can change the shape of but that can go back to their original shape e.g. a squashed piece of sponge. Such explorations can be useful but they contain a hidden problem. This regards the objects that change their shape back to their original form, such as a squashed sponge, or a stretched elastic band.

The problem here is how the objects regain their original shape. It is easy for the children to appreciate that a push from their hand might squash the sponge but they are likely to be confused about how it changes back. Unless they are encouraged to think about this they are likely to offer explanations such as, 'it just goes back', or, 'you aren't pushing it any more, so it won't stay squashed'. In other words, the children don't realize a force is needed to change the object's shape back to what it was before. This is another area where the forces facts are helpful, as, if the children are familiar with them, the notion that the objects just change back can be challenged: if a sponge changes from a squashed shape back to its original shape, then that is changing shape; and changing shape cannot happen without a push, or a pull to effect it. Young children are likely to struggle to come up with suggestions as to where this force comes from but the key thing to help them focus on is that such materials are springy and that it is their springiness that pulls and pushes them back into shape. Some able children may link this to how the particles in the material behave and, if they have a good understanding of materials, they might also appreciate that in examples such as the sponge, the air particles rushing back into the holes will also help push it back into shape, but, at this stage, such sophisticated reasoning is less likely. This is not a problem, so long as the children appreciate that there are forces at work and that they come from the springiness of the material; a fact that they will appreciate more readily if, with appropriate consideration of safety issues, they can play with springs and elastic bands and feel them push and pull back when they are squashed, or stretched. Further refinement can come in key stage 2 and is discussed under the third conceptual step in Chapter 19.

19

Developing a more advanced understanding of forces

Introduction

As was discussed in Chapter 18, forces is one of the most conceptually difficult primary science topics that there is and there is no doubt that the concepts discussed here can be hard for children to get to grips with and that they often have serious misconceptions regarding them. However, with care, it is possible to successfully challenge these misconceptions and have a great deal of fun learning about forces.

Where this topic fits in

It is very important that, before progressing to the learning outlined below, the children are secure in the concepts covered in Chapter 18, especially the basic forces facts, which were summarized under the second conceptual step in that chapter. These will be an important aid in helping the children make the observations which will be necessary to challenge misconceptions and it is vital that both you and the children are secure in understanding them. For your convenience, the forces facts are listed again before the conceptual steps for this chapter.

Both of the conceptual steps in Chapter 18 should be covered in key stage 1. However, the first conceptual step of this chapter also contains much that is accessible to older key stage 1 children. Sometimes it might be desirable to introduce these concepts at that stage and advice on this is given in the relevant section. In fact, I have often found that capable upper key stage 1 children can also cope well with the concepts of step two, though it would normally be best covered in lower key stage 2.

The remaining two conceptual steps should definitely be tackled in key stage 2. If steps two to four are being covered in key stage 2, they contain such a spread of learning that it is advisable to split them into two separate slots. The second conceptual step is relatively straightforward and suitable for covering lower in the key stage; the other two are best left to later in the key stage.

Links to other topics

It is important that the children have a sound grounding in a basic model of particle theory (as discussed in Chapters 8 and 9) if they are to be able to explain some of the forces phenomena that they will encounter.

A context for the topic

The ubiquity of forces means that there are a host of exciting cross-curricular topic themes suitable for learning about them. Two slightly different topics will provide useful examples for this discussion.

The first was a history topic where the children were learning about the Scottish Wars of Independence. This was already an exciting cross-curricular topic but it also proved an ideal context for learning about forces through practical work based on siege warfare. The children had great fun building and testing siege towers, battering rams and trebuchets, which progressed their forces concepts extremely well. This topic provides a good example of where effective science learning can be founded on a topic based in another curricular area.

The second example is one where science formed the lead topic but also opened the way for effective learning in other curricular areas. This was a topic based on rockets. Building rockets powered by a variety of safe propellants was a splendid way to learn about forces and offered great opportunities for investigative science.

The basic forces facts, as summarized in Chapter 18

These key facts are central to an understanding of forces and so are summarized again:

1 That forces are either pushes, or pulls;
2 That forces can only do four things: speed objects up; slow objects down; change the direction of an object's movement; change an object's shape;
3 That the four things listed under point 2 cannot happen without at least one push, or pull to cause them.

Conceptual step one – measuring and quantifying forces

The first steps towards this conceptual step might be made in key stage 1. The concepts are not difficult and many key stage 1 children will be able to carry out simple investigative work that involves their measuring pushes and pulls using non-standard units. This can then be refined and the standardized unit of the Newton introduced in key stage 2.

Eliciting ideas

The best way to elicit the children's ideas regarding measuring forces is to set them practical investigative tasks that require them to measure forces.

Table 19.1 Example learning objectives for developing a more advanced understanding of forces (the first of these learning objectives is suitable for coverage at either upper key stage 1, or lower key stage 2, although I would not introduce the standardized unit of the Newton until lower key stage 2; some able upper key stage 1 children can actually cope well with the second conceptual step but, in general, the final three conceptual steps are best left until later in key stage 2)

Conceptual step	*Example learning objectives*
	Children will learn:
Measuring and quantifying forces.	how to measure forces using gravity buckets and non-standard units; how to measure forces using a variety of push and pull meters, first using non-standard units and then Newtons;
Understanding friction and drag.	that friction is a force that opposes motion between solid objects and to be able to identify common implications of this; that drag is a force that opposes motion between solid objects and liquids and gases and to be able to identify common implications of this regarding air and water resistance;
Understanding balanced forces.	to identify forces in everyday situations and use simple diagrams to illustrate them; that when objects are at rest, the forces acting on them are balanced and that balanced forces can also occur when objects are moving at a constant speed in one direction; how solid objects act like springs and can exert pushes and pulls;
Understanding gravity and falling.	that gravity is a force of attraction between all objects the strength of which depends on how massive the objects are, and how far apart they are; that weight is an example of force and is caused by gravity pulling an object; that, on Earth, it is an object's air resistance that is the primary factor affecting how quickly it falls, not its weight.

For example, when I was working with key stage 1 children examining forces in the context of a topic on explorers, different groups of children had the task of planning explorations to different habitats and one group had to plan an Antarctic expedition. They were set the task of working out how many dogs they needed to pull their sledges. The children could discuss how to test this effectively using real sledges and dogs but the make-believe context of the classroom was more challenging. As is particularly the case for younger children, you must be ready to provide them with extra scaffolding to help their progress when they encounter new learning situations. In this case, the children were provided with model sledges and their loads, a box of different plastic toy dogs, and a gravity bucket. In fact the children quickly worked out how to use the gravity bucket. (Children can be seen using a gravity bucket in a different context in Figure 18.2 in Chapter 18.)

A lower key stage 2 example from the siege warfare context was where the children had to work out how many oxen it took to pull their king's prize bombard on the way to blast the enemy castle. Having done work similar to that discussed above in key stage 1, the children were quick to suggest gravity buckets and they decided to use marbles to represent the pull of the oxen.

Be warned!

Measuring pushes and pulls in contexts such as the above is relatively easy when using very concrete examples such as plastic dogs representing the pull of a real dog, or answering questions such as how many *Compare Bear* explorers it takes to pull a load up a cliff. It can be harder for some children to move to using more abstract non-standard units such as marbles so I always plan to use the concrete examples first. Once the children are used to using for example a *Compare Bear* to represent the pull of one bear explorer, they can much more easily move on to accepting that a marble might represent the pull of one: ox hauling a bombard; Viking portaging his longship; Egyptian dragging a pyramid block etc.

Challenging the children's ideas

It can be seen that there are no tricky misconceptions in the measurement of pushes and pulls but it is important to plan a suitable progression. I aim for the following steps:

1 Introducing measurement using very concrete examples such as plastic dogs to represent the pull of a dog, as discussed in the sledge example.
2 Extending these concrete examples to a variety of contexts. For example, the sledge testing group present their methods and results to the other children who then have a variety of investigations such as testing: how many *Compare Bear* paddlers it takes to move canoes of different loads; how many sacks of supplies a cardboard camel can carry; and how many boxes of stores can be loaded into different model boats.
3 The next step is to help the children move to more abstract non-standard units, such as the example above when marbles were used in a gravity bucket to present the pulling power of oxen.
4 I then find it useful to help the children appreciate that weight is an example of a force. There was discussion in Chapter 18 as to how younger children should be introduced to gravity in a very basic way and how, as part of this, they should explore how objects would push down on their hands when they were holding them (because, since gravity was pulling them, the objects had weight). The activities in the first three steps above can be used to reinforce this, for instance: the crates push the boat lower in the water because they are heavy (they have weight, as gravity is pulling them); or, the marbles in the gravity bucket pull the

Siege warfare makes a great context for the children to learn about forces. The cross-curricular learning opportunities are excellent, not just from a historical point of view but also from design and technology where the children design, build and test their own siege engines, such as this trebuchet.

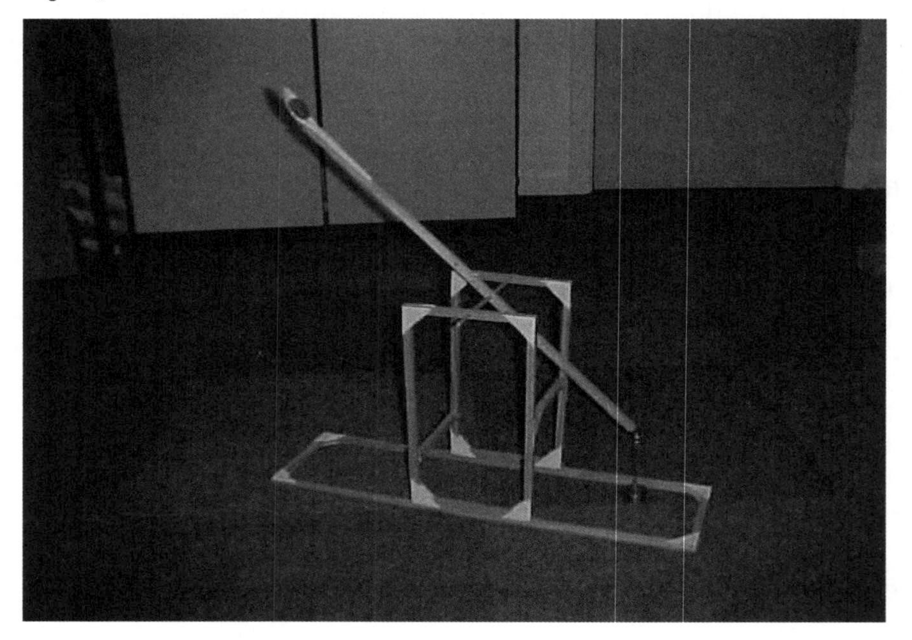

Figure 19.1 Model trebuchet

cannon along because they are heavy (they have weight, as gravity is pulling them). I have then often found it quite natural for children to suggest using standard weights (grams etc.) as a measure of pushes and pulls, for example as the counterweight for a model trebuchet as in Figure 19.1. This helps them further appreciate that weight is a force, though you should be alert to the fact that many will at first think of the weight as the cause of the pull, rather than the pull itself.

5 Finally the children can be introduced to the standard unit for measuring forces, the Newton, and how this relates to measures of mass, such as the kilogram. At this point, they can also be introduced to a variety of push and pull meters, such as 'bathroom' type scales and spring balances. It's best for work in science, if these are calibrated in Newtons.

Conceptual step two – understanding friction and drag

It is not difficult to introduce some excellent investigative work when learning about friction and drag. The basic concepts are not especially difficult but these areas are also likely to raise questions where the children may have deep misconceptions, so it is worth thinking carefully about how they are introduced.

Eliciting ideas

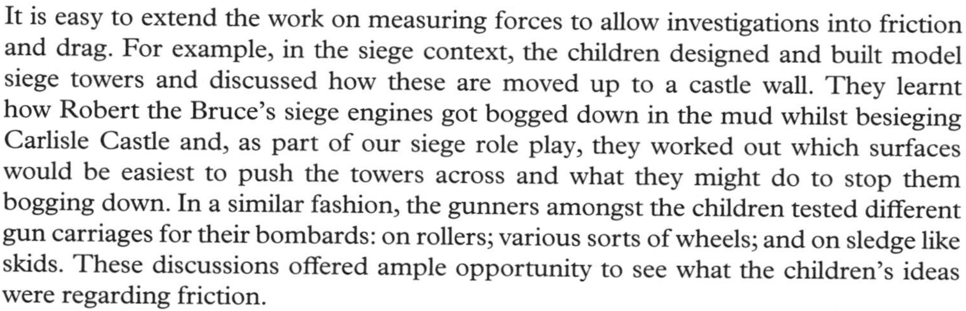

It is easy to extend the work on measuring forces to allow investigations into friction and drag. For example, in the siege context, the children designed and built model siege towers and discussed how these are moved up to a castle wall. They learnt how Robert the Bruce's siege engines got bogged down in the mud whilst besieging Carlisle Castle and, as part of our siege role play, they worked out which surfaces would be easiest to push the towers across and what they might do to stop them bogging down. In a similar fashion, the gunners amongst the children tested different gun carriages for their bombards: on rollers; various sorts of wheels; and on sledge like skids. These discussions offered ample opportunity to see what the children's ideas were regarding friction.

Such a context also allows opportunities to explore the children's ideas on drag. The children learn that moving armies and all their supplies in medieval times could be very difficult due to the poor nature of the roads. This allows for a debate about whether, where possible, it would be easier to move heavy loads such as the components for the siege engines by water, rather than land. This allows the children to employ their techniques for measuring how much force it takes to move loads in a comparison between moving them on carts or on barges. Further opportunities to elicit their ideas on drag are offered by setting them the task of finding out which shape of barge was easiest to move through the water.

In this medieval context, it is a bit harder to find ways to explore drag in the form of air resistance but even this is not impossible. When testing how to maximize the range of their trebuchets, the children discuss how the shape of the missile might make a difference to how well it flies through the air and this then makes it possible to find out the children's ideas on streamlining and air resistance.

Be warned!

Children are usually quick to appreciate that rough surfaces have more friction than smooth surfaces and that lubricants will reduce friction but they can still be confused about some aspects of the concept. Sometimes I have found children who believe that smooth surfaces have no friction at all. More common are children who, in contexts such as trying to find the easiest way to move a siege tower, will claim that 'all friction is bad', not realizing that some friction is necessary for the tower's wheels to grip the ground for it to be able to be moved, steered, and braked.

Drag in liquids, such as water resistance, similarly usually doesn't present too many difficulties for the children and they are quick to appreciate how more streamlined shapes are easier to move through the water.

I used to find that drag in gases caused more confusion but soon realized that this had more to do with the children's lack of understanding of gases as corporeal materials. However, I found that once I ensured that the children had had a proper grounding in materials, including a simple particle theory of matter (as discussed in Chapters 6 and 8), these problems largely disappeared.

Work in this area may also uncover a more complex, deeper misconception that

that can be challenged through exploring friction. This is the common belief that forces are not needed to ultimately stop a moving object. Children will happily accept that, for example, brakes may push against a wheel to stop a model cart more quickly but yet, if asked why the model soon stops after it is given a push across the carpet, they will often say something like, 'you didn't push it hard enough', or, 'your push has worn off'.

Challenging the children's ideas

The problem solving activities discussed above as a means of eliciting the children's ideas naturally lead to investigations that allow the children to test their ideas and to have the appropriate scientific ideas consolidated.

The siege tower and gun carriage examples are perfect for the children to test different surfaces and how their friction varies. Likewise, the use of 'boat hauling' apparatus (as shown in Chapter 18) allow the children to test their ideas on water resistance.

Air resistance is harder to allow the children to test in this context. The children test different missile shapes with their trebuchets and discover that the effect of air resistance on these was often less than the natural error in results due to such factors as judging where the missiles impacted and slight variations in the swing of the throwing arm. However, the great thing about encountering problems when investigating with the children is that, so long as they are well motivated, this leads naturally to excellent thinking skills and problem solving development. In this case, the children design their own air resistance test based on dropping identical weights inside different cardboard shapes.

I have found that investigations such as these allow the children to connect with the necessary concepts very well and that little else is required apart from consolidation work such as helping the children find ways to summarize their knowledge in diagrams. Acting out what happens can also be useful, especially for drag, where the children can act out different shapes pushing their way through liquid, or gas particles.

Challenging the misconception that forces 'wear off' is harder but this is the ideal time to discuss it with the children. The key point is to focus on is how it is the friction and/or drag that slows and stops an object moving. The investigation contexts will help the children realize this and it is also important to refer back to the basic forces facts that were restated at the beginning of this chapter. These will help the children in that they show that if an object is to slow down, there must have been a force that did this; it can't be that a moving force has stopped, or worn off. Where is the force that has slowed the object? It is the push from the particles of the materials the object is moving across and through: the ground, water, or air. This can lead to considering what would happen if there was no push from friction or drag to stop an object. If you pushed an object in such circumstances, it would never stop. However, this concept might be rather too abstract for typical lower key stage 2 children and might be better left until they revisit forces later. (It is discussed under conceptual step four.)

Conceptual step three – understanding balanced forces

This can be one of the trickier areas of forces, though, once again, if the children have followed the conceptual steps outlined before, I have found that their progress is much more trouble free. Depending on your learners, you may wish to leave exploring these concepts to later in key stage 2. I have found that younger key stage 2 children often cope well with the idea of balanced forces regarding stationary objects but that they can struggle more when it comes to balanced forces regarding moving objects.

Eliciting ideas

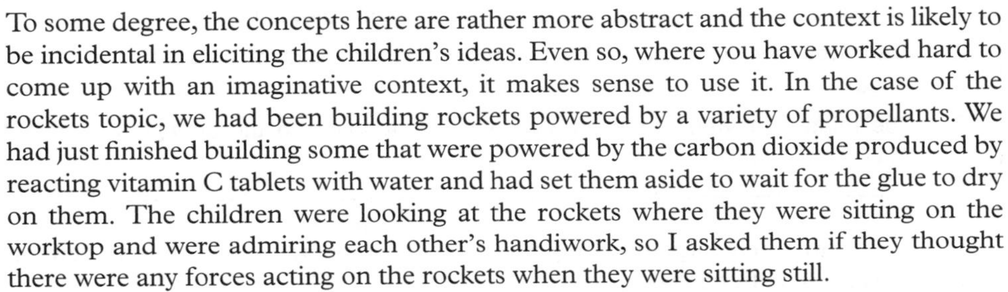

To some degree, the concepts here are rather more abstract and the context is likely to be incidental in eliciting the children's ideas. Even so, where you have worked hard to come up with an imaginative context, it makes sense to use it. In the case of the rockets topic, we had been building rockets powered by a variety of propellants. We had just finished building some that were powered by the carbon dioxide produced by reacting vitamin C tablets with water and had set them aside to wait for the glue to dry on them. The children were looking at the rockets where they were sitting on the worktop and were admiring each other's handiwork, so I asked them if they thought there were any forces acting on the rockets when they were sitting still.

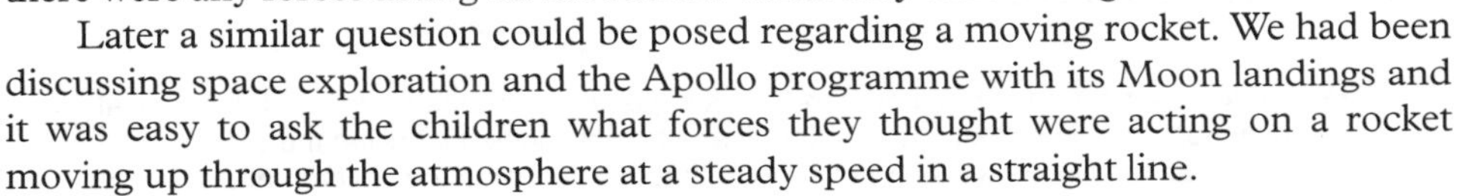

Later a similar question could be posed regarding a moving rocket. We had been discussing space exploration and the Apollo programme with its Moon landings and it was easy to ask the children what forces they thought were acting on a rocket moving up through the atmosphere at a steady speed in a straight line.

Be warned!

It is a classic misconception that children believe that stationary objects have no forces acting on them. However, I have found that children who have engaged with the sorts of learning activities outlined above, and in Chapter 18, rarely believe this. In the example of the rocket, many children told me that 'gravity is still pulling the rocket', whilst others could even tell me things such as that 'the rocket is pushing the worktop because gravity makes it heavy'. This still doesn't mean that the children understand the forces acting on the rocket are balanced. Very few children will initially have any comprehension that the worktop is actually pushing back on the rocket. This is hardly surprising as it is far from obvious how the worktop can actually push anything. Instead the children will be most likely to hold the idea that the worktop's role is a passive one and they will tell you things such as, 'the worktop stops the rocket sinking' or, 'the worktop gets in the way, so the rocket can't fall anywhere' or, 'the worktop supports the rocket'.

You are likely to find similar confusion regarding balanced forces on moving objects. In all my experience, I have only ever found three primary school children who, when initially asked, realized that if an object is moving at a steady speed in a straight line, the forces acting on it must be balanced, or, as one of the children put it to me, 'cancel each other out'. Most children, realizing that friction and drag have to be overcome, naturally jump to the erroneous conclusion that whatever push or pull is responsible for the object moving, must be greater than the opposing forces of friction and drag.

Hint!

At this stage it is useful to introduce the children to drawing simple forces diagrams. This will give you another means of eliciting the children's ideas and will also be useful when it comes to help the children summarize and consolidate the correct ideas. The diagrams need not be complex and it is a good idea to limit the forces shown on the diagram to the ones relevant to the key concepts being discussed. There are three simple rules for showing the forces:

1. Arrows are used to show the forces and the direction they act in.
2. The arrow should show, as best as possible, the point of application of the force.
3. The relative lengths of the arrows should correspond to the strengths of the forces.

Figure 19.2 shows such diagrams being used to represent both a child's misconception and the correct scientific ideas.

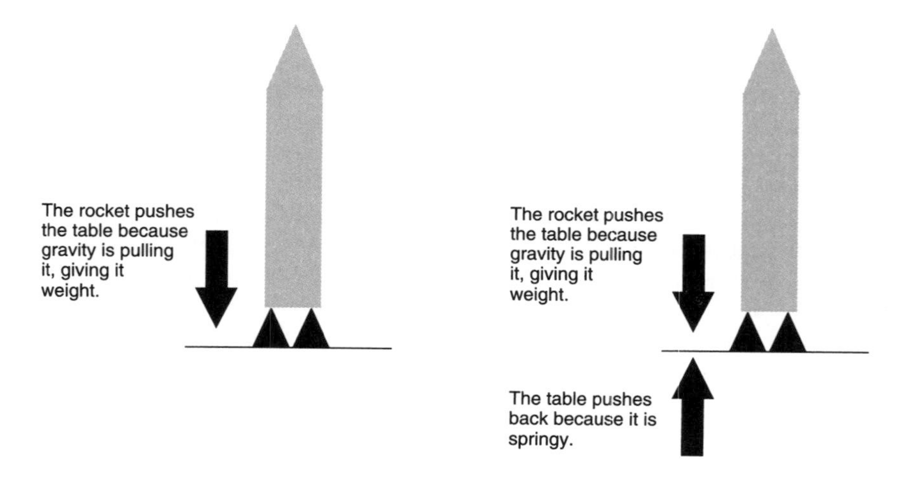
Example forces diagrams for a model rocket sitting on a worktop, showing both the typical misconception and the correct ideas. (Note some books suggest labelling the downward force simply as gravity. However, I have found that this confuses the children and that it is helpful to focus on the concept that, because gravity is pulling the rocket, it has weight, and so pushes down on the table, which, in reaction, pushes back. This will also help ensure less confusion for the children when they come to encounter Newton's laws of motion in later key stages.)
The rocket pushes the table because gravity is pulling it, giving it weight.
The rocket pushes the table because gravity is pulling it, giving it weight.
The table pushes back because it is springy.
The common misconception
The correct scientific idea: the forces are equal and therefore balanced.

Figure 19.2 Example forces diagrams for a model rocket sitting on a worktop

Challenging the children's ideas

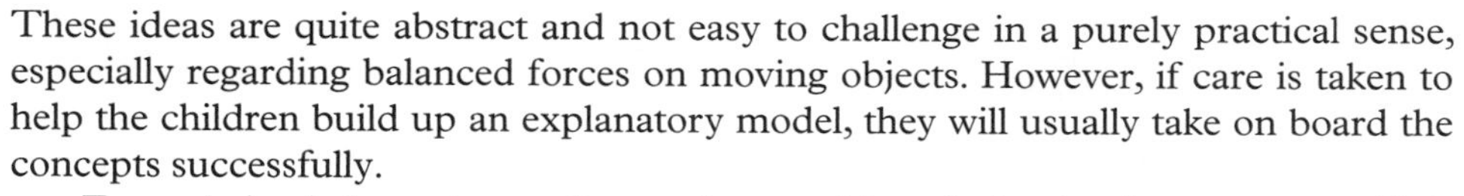

These ideas are quite abstract and not easy to challenge in a purely practical sense, especially regarding balanced forces on moving objects. However, if care is taken to help the children build up an explanatory model, they will usually take on board the concepts successfully.

Forces being balanced regarding stationary objects is the simpler concept and it is best to start there. The key problem is to allow the children to appreciate how a worktop, the ground etc. can push back and this builds on the concept of springiness, as introduced in the last chapter. In the rocket example I began by setting the children the problem of making a paper bridge out of two sheets of A4 paper that could support their rocket a minimum of ten centimetres above the table. After some experimenting, they all duly did this but the key conceptual point was that they noticed that: when the rocket was set on the bridge, the bridge bowed; and, when the rocket was lifted off, the bridge sprang back. In other words, the bridge acted as a spring. The children were challenged to find similar examples and came up with an imaginative variety, including getting me to stand on a plank placed between two benches in the school hall! The challenge was then to explain why some materials were springy. I was very pleased that, since the children had been introduced to the models of materials and their properties discussed in Chapters 6–9, the children quickly suggested that the particles in springy materials must be able to squeeze and stretch. A couple of very able children even linked this to their work in electricity and suggested that this was, 'because of the pluses and minuses in the particles attracting and repelling'; excellent conceptual reasoning! From this it was a comparatively easy step to realizing that all solids (and gases) are springy, even if only at a particle level and that this is what allows a worktop to push back on a rocket set on it; the ground to support us by pushing back on us; or a ball to bounce off a wall as the particles in each compress and spring back. These concepts were then made more accessible, with due care, by the children pretending to be particles and acting them out, as shown in Figure 19.3.

Be warned!

Whilst this sort of model explains how a solid worktop can push back against the weight of an object set on it, it does not explain why some objects float in liquids.

Balanced forces regarding floating objects work in a quite different way, as liquids are not compressible, or springy. It is possible to have primary school children work up to an excellent practical understanding of floating and sinking but it takes quite a lot of time and I usually consider it best left until key stage 3. However, curious children may well raise the issue and it is worthwhile considering how to address this.

What you should certainly avoid is the kind of feeble explanation that I remember telling children when I was a newly qualified teacher! I can clearly recollect some children in my first ever class asking me why some things floated. In response I launched into a (bad) explanation of density, which I now realize was of no help to the children at all.

My primary school explanation of floating focuses on helping the children appreciate the following points:

1. When an object floats in water, it pushes some water up out of the way.
2. This means the water level rises.
3. Gravity tries to pull the water that has been pushed up back into the space from which it has been pushed.
4. The pull of gravity on the lifted water makes the water push against the object and it is this push that makes the object float.
5. If an object pushes out of the way a volume of water that weighs the same as it does, it floats; if the volume of water the object pushes out of the way weighs less than it does, it sinks.

It is possible to let the children explore some of these phenomena quickly and easily, such as observing water level rise when an object floats in it but practically demonstrating the exact relationship of point 5 is quite difficult in primary school.

Some students act as wall particles while others lean on them. The wall particles are compressed and the force of their springiness supports the leaning students.

Figure 19.3 Students modelling how a wall can push back against someone leaning on it

Here the child has incorrectly shown the force from the rocket motors as being greater than the combined force of gravity and air resistance. This would mean that the rocket was accelerating, rather than moving at a steady speed.

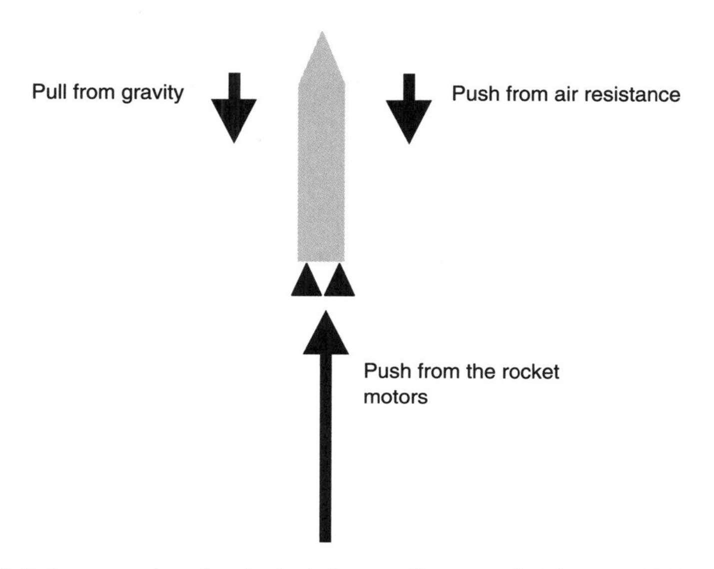

Figure 19.4 An example of a typical forces diagram showing a child's misconception that imbalanced forces are needed to keep an object moving at a steady speed in a straight line

The concept of balanced forces on moving objects is trickier and not amenable to practical elucidation. Here the value of the basic forces facts developed in Chapter 18 is once again demonstrated.

When asked to draw a forces diagram for the rocket moving up through the atmosphere at steady speed in a straight line, most children drew a diagram such as Figure 19.4.

This misconception is entirely understandable and the best way to challenge it is to appeal to the forces facts and to ask the children a series of questions based on them as follows:

1 Is the rocket speeding up? no
2 Is the rocket slowing down? no
3 Is it changing direction? no
4 Is it changing shape? no

Therefore none of the four things that forces can do are in evidence but . . .

5 Are forces acting on the rocket? yes: push from the engine; gravity is pulling it; and there is drag from the air.

Therefore, if forces are acting on the rocket but it is neither speeding up; slowing down; changing direction; nor changing shape – the forces must be balanced, as in Figure 19.5.

Here the force of the rocket motors equals the combined forces of gravity and air resistance.

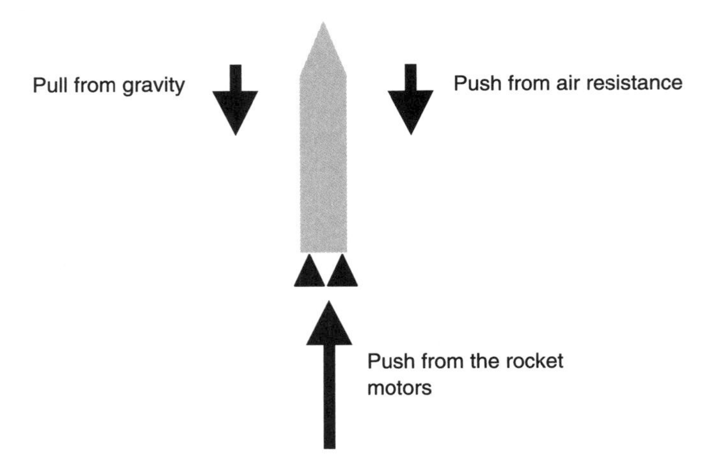

Figure 19.5 Forces diagram showing the correct forces acting on a rocket ascending through the atmosphere at a steady speed in a straight line

This is rather abstract reasoning but I have found that if the children have had the more concrete learning experiences leading up to this, that they accept it readily.

Conceptual step four – understanding gravity and falling

This area offers some splendid opportunities to develop the children's thinking skills but it is difficult and misconceptions are common.

Eliciting ideas

The rocket topic offered obvious opportunities for finding out the children's thinking on gravity and discussions are prompted via several means.

The children built model rockets powered by water; by carbon dioxide; and in collaboration with a nearby aerospace museum, by real solid fuel motors. This offered the chance to discuss how high we can get the various rockets to go and as to why our model rockets always fell back to Earth.

Setting the children the task of predicting how quickly which of three rockets of different sizes and weights would fall also provide insight into their thinking.

Finally, we had great fun (and excellent cross-curricular learning opportunities) through designing parachute recovery systems for the rockets, which then promoted further discussion as to why parachutes slow down the rockets.

Be warned!

Mention was made in the previous chapter of how many young children believe that no force is necessary to make objects fall. Hopefully, if the children have engaged with the learning that has been discussed, they will now realize that this is not the case but this is not to say that they will understand gravity and care must be taken in using elicitation activities to check exactly what their thinking is as there are many common problems.

Children have often told me that gravity is only found on Earth, an understandable enough confusion as they often have images of astronauts floating in zero gravity in mind, which they then extend to anywhere in outer space, such as one child who once emphatically put it to me that, 'Earth is where gravity is. There's none in space: not on stars, moons, planets, comets, or anywhere like that'. The confusion is often made worse by many science books only ever referring to gravity as, 'the Earth's pull'.

Other confusions can be of a simple, factual nature, such as thinking gravity is a push, rather than a pull, or less obvious; for example, I remember being initially surprised when I realized that quite a few of the children I was working with thought gravity only acted on solids. (I still find a significant percentage of student teachers do not realize that it is gravity that keeps our atmosphere in place!)

More problematic can be the children's beliefs as to what actually causes gravity. I have encountered three common explanations (which, incidentally, I have found as prevalent amongst adults, as amongst children).

The first is that gravity is caused by the Earth spinning. Some children have told me this but I have found this misconception more common amongst adults who are often getting confused with their notions of centripetal and centrifugal forces.

I have also found that children regularly associate gravity with the Earth's magnetic field. Since both gravity and magnetism are non-contact forces acting at a distance, this is not surprising. As is often the case, the children's reasoning is sophisticated, if misplaced. For example, one girl told me, 'there is a middle bit in the Earth called the core that's iron and like a giant magnet. It's so powerful that it doesn't just pull stuff with iron in it but it can pull anything. Heavy things have more iron in them and so they get pulled more and fall faster'. Even acknowledging the inconsistency, and the misconceptions, this sort of reasoning is very impressive. (Perhaps it was also no surprise that this child's imaginative writing was excellent!)

However, by far the most common explanation for gravity amongst children is that it is caused by the air. Often, this has elements of confusion with air pressure but, once again, the reasoning behind the misconception is often sophisticated. A dialogue I had with one year five boy is worth reporting in full to illustrate how even the most able children can become confused.

Boy: It's gravity that causes things to fall.
Teacher: What causes gravity?
Boy: Have you ever noticed that spacemen wear helmets?
Teacher: Yes.
Boy: That's it.
Teacher: Sorry, I'm not sure I know what you mean.
Boy: Why do they need helmets?

Teacher:	(realizing what the child is thinking) To breathe?
Boy:	Exactly! There's no air in space.
Teacher:	What do you mean?
Boy:	There's no air. That's why they just float about in space. It's air that causes gravity.

Never before, nor since, have I encountered a child who has questioned me like this. He was a born educator; he just needed to improve his subject knowledge a bit!

Finally, regarding falling itself, you are almost certainly going to encounter the extremely common misconception that weight is the primary factor defining how fast objects fall and the corresponding lack of appreciation of the importance of drag in determining an object's speed of descent.

Challenging the children's ideas

A range of strategies needs to be employed to challenge these misconceptions.

The context of rockets and spaceflight allowed a natural challenge to the ideas that gravity is only found on Earth and the often related notion that gravity is caused by air.

For example, it was relatively straightforward to challenge the thinking of the boy in the example quoted above. All that was needed was to show him video footage of astronauts floating around inside a space station. Within seconds the boy, without prompting, said, 'they've not got helmets on'. He had realized that there was no gravity inside the space station but that, since the astronauts had no helmets on, there must be air and that therefore air could not be the cause of gravity.

Similar strategies can be used to challenge the belief that gravity is only found on Earth. If the children watch footage of astronauts walking and bouncing on the Moon, they can be asked why the astronauts fall back to the surface when they bounce. Here the value of the forces facts first introduced in Chapter 18 can be seen once more: if the bouncing astronaut slows and changes direction (falls back to the surface) there must have been a force that caused this; in other words, gravity. It is then relatively straightforward to introduce the key concepts that gravity is a force of attraction between all objects; the strength of which depends on how massive the objects are, and how far apart they are; and that the force acts as a pull towards the centre of the objects. Gravity is such a relatively weak force that we only notice the effects from very massive objects such as planets, stars and moons. This is a very abstract concept but, if the children have been encouraged to think through the learning discussed so far, I have found that they accept the concept without difficulty. Many curious children will want to know exactly why this works but it is another very useful example of where even expert scientists are still trying to understand exactly what is happening. Many scientists predict that gravity particles will be discovered that work in a similar way to light particles and there are several experiments attempting to detect them. However, because of the weak nature of gravity, this is very difficult and, so far, none have been detected.

Hint!

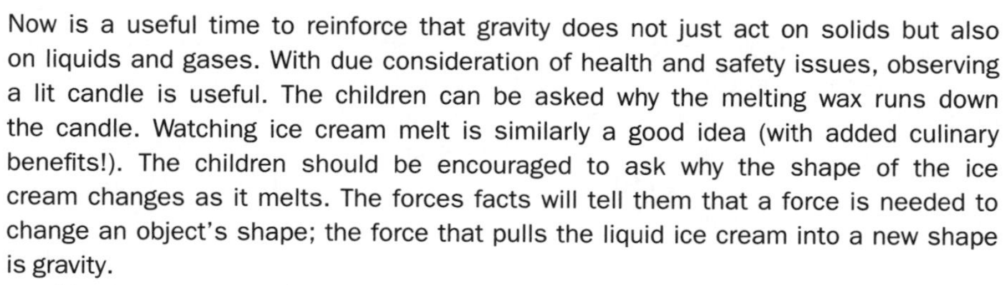

Now is a useful time to reinforce that gravity does not just act on solids but also on liquids and gases. With due consideration of health and safety issues, observing a lit candle is useful. The children can be asked why the melting wax runs down the candle. Watching ice cream melt is similarly a good idea (with added culinary benefits!). The children should be encouraged to ask why the shape of the ice cream changes as it melts. The forces facts will tell them that a force is needed to change an object's shape; the force that pulls the liquid ice cream into a new shape is gravity.

You may link these observations to explaining why we have tides. The exact mechanics behind the tides may be a bit beyond typical primary school children but they can readily grasp that they are due to the Moon's (and, to some extent, the Sun's) gravity pulling the liquid seas about on the surface of the Earth.

Gases present a problem, so far as practical demonstration is concerned as the dense, visible gases that children might watch being pulled down by gravity are much too dangerous for use in a primary school context. However, if the children have a good appreciation of gases as materials, all is not lost. I used to get children to weigh two of my identical SCUBA cylinders: one empty and one full of compressed air. This readily demonstrates that gases have weight and the children should now be appreciating that weight is a result of gravity's pull. (The cylinders and the gas in them would be weightless if we took them out into space, away from planets, moons etc.) This can then be linked to why the atmosphere does not just float off into space; luckily for us, the pull of gravity keeps it around the Earth.

At this point it is important to introduce the distinction between weight and mass. The children now usually readily appreciate that weight is due to the pull of gravity and, as a force, should be measured in newtons (which they can easily do using a variety of push and pull meters). They will also appreciate that objects' weight can change depending on the strength of gravity acting on them: they will have seen, for example, images of astronauts floating weightless in zero gravity conditions; and should now be able to make the link between astronauts being able to bounce about on the Moon and the fact that there is less gravity on the Moon. The concept of mass is a bit harder. I aim to help the children use their understanding of materials to appreciate that mass links to the particles that are in a material. Even if the material is weightless, the particles are still there. They will then see that, although if we stay on Earth we usually have no need to distinguish between mass in kilograms and weight in newtons, if we venture away from the Earth, the distinction becomes more important. For example, astronauts on the Moon will have the same particles in them as they would have on Earth and therefore the same mass, but they will only weigh about a sixth of what they do on Earth, due to the Moon having less mass than the Earth, and therefore a weaker pull of gravity. Useful activities to help reinforce these concepts include the children designing a spreadsheet to calculate their weight on different planets in the solar system, or converting recipes for use on other planets and moons.

Be warned!

The children may get some surprises when carrying out exercises such as those just discussed. For example, Jupiter's pull of gravity, whilst over twice that of Earth, is much less than its relative size might suggest. I often deliberately frame such exercises so that the children make such observations. This allows them to begin to think about what, at key stage 3, they will learn as density. At this level, I do not worry about the children learning the technical aspects of density but, once again, they can use their knowledge of materials to help their thinking. They can research what the different planets are made of and find out, for instance, that the Earth contains lots of molten iron and rock but that Jupiter is mostly made of ice (frozen gases of various types). The children can then make some simple comparisons such as comparing a kilogram of iron to a kilogram of ice. They will see that, although they weigh the same, the volume of ice is much greater. The particles are much more squeezed up in the iron, than in the ice. This is a simplification but it helps the children understand their observations. Jupiter is much bigger than Earth but because it is made mostly of ice, its mass is less than you might expect for its size.

Hint!

The children's Earth-bound experience will never allow them to appreciate a situation free from the effects of friction, drag and gravity and so, as was discussed under conceptual step two, they often think that moving objects slow, or fall, because the effects of forces that initially moved them wear off. The learning situations already discussed will hopefully have gone a long way to challenging this misconception but now is a good time, especially in a rocket, or space, topic to return to this difficulty.

For example, I normally ask the children to consider a rocket blasting along in space, far away from stars, planets, moons etc. I ask them to think about what would happen if the rocket motor was turned off. Despite all the challenges discussed, some children will predict that the rocket will slow down and will have to be encouraged to see that, since there are no forces to slow the rocket (no friction, drag, or gravity) it will keep moving at a steady speed in the direction it was going. This can then lead to exploration and discussion of how astronauts only use the rocket motor to: accelerate to the speed they wish; to change direction; and to slow down. If they are not doing any of these things, the motor is turned off.

The final area to concentrate on is falling objects. The elicitation exercise referred to surprised all the children who had confidently (and predictably) stated that the heaviest model rocket would fall the fastest. They were initially baffled as to how objects of different weights could fall at the same speed but this led nicely to excellent investigative work to establish what affects the rate at which an object falls. These investigations quickly confirmed that weight is not the crucial factor in determining how quickly the objects fall. The tests showed that drag (air resistance) plays a significant role in affecting the speed.

Hint!

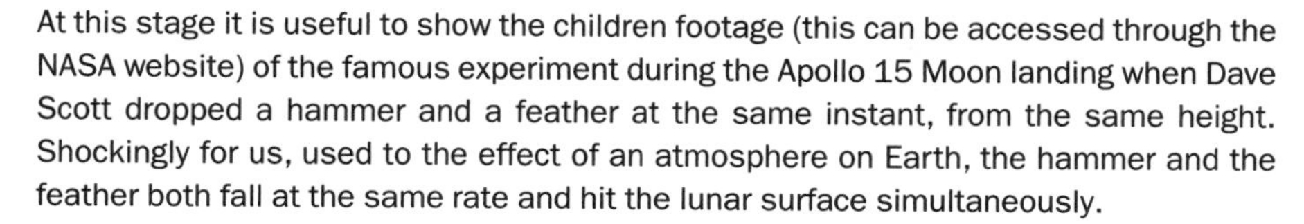

At this stage it is useful to show the children footage (this can be accessed through the NASA website) of the famous experiment during the Apollo 15 Moon landing when Dave Scott dropped a hammer and a feather at the same instant, from the same height. Shockingly for us, used to the effect of an atmosphere on Earth, the hammer and the feather both fall at the same rate and hit the lunar surface simultaneously.

Of course, if the children go on to investigate the effect, for example, of suspending different loads under parachutes, or putting different weights on paper spinners, they will discover that weight can make a difference! I do not think that at this level it is advisable to be going into the intricacies of terminal velocity, still less drag coefficients, and I am content if the children realize that, in an atmosphere, drag plays the major role in determining how quickly an object falls.

Hint!

Controlling variables can be tricky for children when investigating what affects the rate at which objects fall. There are three areas where there may be problems, all of which can be solved with a little thought.

Firstly, I often find that children suggest changing both shape and surface area and then get a bit confused when they realize that the two are interlinked.

Secondly, and immediately following on from the above is the fact that the children often struggle to think of a way to control the variables so that they are changing just shape, or surface area and not other variables, for example the weight of the object.

So far as the second problem is concerned, I steer the children towards paper as a suitable material as it easy to use identical sheets of paper as means of controlling variables whilst changing their shape/surface area. Sometimes using a malleable material such as modelling clay is suggested as a means of addressing the problem but it is very unlikely that changing the shape of a given lump of such a material will make sufficient difference for the children to be able to measure, which may lead to the children missing important concepts.

Using paper provides an opportunity for the children to get to grips with the first difficulty. For example, the paper can be kept as a rectangle but folded in half, quarter etc. to change its surface area, which will clearly demonstrate to the children the importance of surface area. Paper also allows the children some scope in investigating changing shape, whilst keeping surface area the same. For example, the children can compare dropping a flat piece of paper, with an identical piece, which is creased so that it falls in a V shape. In this example the surface area remains identical but the change in shape allows the air to be pushed out of the way more efficiently, which will allow the sheet to fall faster. The key point here is that the amount of drag is affected by both shape and surface area.

The third difficulty also involves controlling variables. The children are sure to wish to test how the weight of an object might affect its rate of descent but often find it hard to think of how to change only the weight of the objects they are dropping. Here, the best strategy is to use identical hollow containers that can be filled with varying amounts of material to change their weight; for example, plastic pop bottles filled with differing amounts of water.

Be warned!

Care needs to be taken when arranging tests of how fast objects fall. In primary school contexts there are no viable alternatives to judging how long it takes something to fall by eye, either comparing objects dropped simultaneously, or timing how long it takes for an object to drop. This can cause difficulties if the factors of error are close to the difference in dropping times. One strategy is to encourage children to see the value in such circumstances of changing their independent variable in quite large steps. For example, if testing the effect of changing rotor diameter on a paper spinner, there is likely to be little value in reducing the diameter of the rotors half a centimetre at a time; 5 centimetre steps might well be more effective at showing the differences. Of course this is just good practice and a science skill the children should be learning anyway.

Trickier can be the problems when the children are comparing the falling of two objects released simultaneously; for example, a small pop bottle full of water, and one only half full. Both bottles will fall at the same rate but it may appear that one falls faster than the other due to: either validity errors in not dropping both at exactly the same height and instant; or accuracy errors in judging which hits the ground first.

There can also be a problem if some children allow their preconceptions to influence how they perceive the results. For example, I have worked with children who, convinced that heavy objects should fall faster, swore that the full pop bottle was hitting the ground ahead of the half empty one, even when their peers were disagreeing. Once again, there is excellent scope for developing science skills here. The children can be encouraged to find more accurate means of judging when the bottles hit, such as the measurer lying at ground level, directly in front of the drop zone (though not too close!). They can also attempt to address validity issues by having a marker for the drop height and perhaps counting down to the release. Reliability can also be invoked. For example, one able but squabbling class of year sixes I worked with decided they would drop the bottles a hundred times, as they couldn't always agree if there had been a difference between the impact times of the bottles. They compared the results of different observers and decided that, statistically speaking, there was no difference in the time taken for the bottles to drop. It kept them gainfully occupied for quite a while, too!

20

Developing an initial understanding of life processes and living things

Introduction

Conceptually speaking, compared to many of the topics discussed, living things and life processes is a relatively straightforward area of primary science and most teachers are far more confident with it than with topics such as forces or electricity. None the less, some aspects regularly cause problems and these are highlighted here.

Where this topic fits in

All the learning discussed in this chapter is suitable for key stage 1 children. Although only the trickiest aspects are discussed here, the overall learning for the topic represents a very considerable spread and it is likely that at least two units of learning will be required for adequate coverage within the key stage. One unit should focus on the children as living organisms and the similarities and differences between them and other living things. One will also be needed to examine living things in the wider context of their habitats. In addition, there may be scope for some extra mini-topics focusing on, for example, specific aspects of the children's life processes in relation to healthy living; and a short unit focusing specifically on plants might also be useful, as they represent life processes at their furthest removed from the children's own experience.

Since the concepts involved are not particularly complex, the learning can be approached even early in key stage 1.

Links to other topics

As always science skills should be at the heart of the children's learning but, in terms of subject knowledge, this topic tends to stand alone.

A context for the topic

Children tend to be fascinated by living things, which makes this an easy topic from a motivational point of view. This means that it is possible to run the topic without any extra theme, or context. Even so, it is important to ensure the children have a particular relevance to their learning, as discussed in Chapter 4. This means that, even when I am planning learning without the sort of elaborate theme represented by, for example, the ice trolls discussed in Chapter 10, I always try to give the children a concrete reason for their work. Some examples of these are given following.

The first work undertaken in this topic will focus on the children's own life processes and then move to comparing these with other living organisms. I have always found the children enjoy such work but added motivation and improved learning will result if the children are given a particular focus for their task. For example, when setting the learning early in a year, I have found a topic of pets works very well. Not only does this provide a useful framework for the learning, but I have found that it provides added focus and motivational excitement if the learning is used to find out what living things can be kept in the classroom and how best to look after them. Keeping living things in the classroom is not only useful for the children's learning in science but also helps to develop responsibility and, I have found, the children's confidence and self-esteem. Of course, due attention must be paid to health and safety issues but, in addition to lots of plants, I used to successfully keep as classroom pets: coldwater fish; tropical fish; giant African snails; various species of stick insects; and a garter snake.

When it comes to leaning about how living things are part of larger habitats, I use several approaches. Sometimes we are explorers finding out about new areas. The science learning can then be combined with the children's first steps in mapping, with learning how to take care of ourselves away from the class and with lots of good language and story work. Wherever possible though, I try to find extra reasons for the children to carry out their work. For example, they might be gathering information with a view to being able to establish, or improve, a wildlife area in the school grounds; such as one year two class that I remember playing a major role in providing suggestions for a new school pond. Also, even at key stage 1, it may be possible for children to contribute to actual scientific surveys, if the required information is accessible to them; amongst other things, I have had key stage 1 children gather information on tree flowering times and ladybirds. This sort of work is also an excellent way to develop the children's science skills and it is always worth checking with local environmental organizations for ideas.

Giving the children a reason to carry out their work also helps with the shorter topics mentioned. For example, rather than just learning about plants, why not set the children the challenge of growing lots of different plants to sell at the school fair? Such a project provides an excellent context for the science learning but also with many useful cross-curricular links.

Even a potentially more mundane topic, such as healthy living, can be enlivened with a creative context. I had one class who really enjoyed helping a group of friendly, though clueless monsters (with life processes that paralleled the children's own) learn how to live more healthily. A friendly local doctor even provided monster diagnoses, much to the children's delight.

Table 20.1 Example learning objectives for developing an initial understanding of life processes and living organisms (all of these conceptual steps are suitable for coverage at key stage 1)

Conceptual step	*Example learning objectives*
Appreciating that humans are living organisms and understanding their major life processes.	Children will learn: that humans are living organisms; that humans have life processes including moving, having a life cycle, having senses, growth, needing oxygen, needing food, and getting rid of waste products; that human bodies are specialized in various ways to perform the above functions;
Extending this understanding of life processes to: first other vertebrates; then invertebrates; and finally to plants.	to categorize entities as living or non-living; that living things share the life processes identified for humans under conceptual step one;
Understanding how living organisms are adapted to different habitats.	how living organisms' life processes are adapted to allow them to live in particular habitats;
Developing simple taxonomic skills.	to categorize living things according to their observable properties; to assign living organisms to the taxonomic categories of vertebrates (mammals, birds, reptiles, amphibians, fish), invertebrates, and plants.

Conceptual step one – appreciating that humans are living organisms and understanding their major life processes

Since young children are naturally egocentric, the best place to begin examining living things is with themselves.

Eliciting ideas

The children will usually have no doubt that they are alive but a useful elicitation exercise is simply to ask them what it is about them that makes them alive. This will not only give you insight into the children's ideas but it will also produce a useful checklist of ideas that can later be tested against other things that the children believe to be alive (or not alive) which will help the children in establishing an understanding of the characteristics of living things.

Be warned!

Few children at this stage will produce a list that corresponds to the biological checklist of the characteristics of living things, sometimes summarized as *MRS GREN*: movement;

reproduction; sensing; growth; respiration; excretion; and nutrition. In any event, in their technical form, some of these are likely to be rather beyond the children at this stage. The children are likely to zero in on some technical aspects such as movement, growth and eating, but the others may not occur to them at first. On the other hand, they are likely to list many features that are not technically characteristics of living things. For example, in my experience, talking, wearing clothes, and having blood, are all commonly stated by children as evidence that they are alive.

Challenging the children's ideas

At this stage I do not worry about correcting all the children's ideas. Some of the more obviously spurious can be challenged by means of simple questions. For example, you might ask the children if they can think of anything that is alive but that doesn't wear clothes. More complex, though technically incorrect suggestions, such as living things having blood, can remain on the class's provisional list of characteristics for further discussion under conceptual step two. What I do concentrate on is supporting the children in ensuring that the following technical characteristics are on their provisional list of the characteristics of living things. Sometimes the technical definition is too complex and so the list is somewhat simplified.

1. Movement. As has already been noted, the children are likely to mention this without prompting.
2. Make more of themselves. For example, humans have babies; birds lay eggs from which hatch chicks; oak trees grow from acorns that are made by their flowers etc. It is also very useful to examine different life cycles as this lays a foundation for challenging the misconception that life can spontaneously generate.
3. Have senses. Here there is scope for simple activities that introduce the children to their own five senses.
4. Grow. Again, the children are unlikely to have difficulty with this.
5. Need oxygen, or air, depending on the children's experience of materials. The biological definition of respiration is too complex at this stage but children can appreciate this simplification and will often list breathing as a feature of their being alive.
6. Need food. Sometimes the children will add water, or 'things to drink' to this, which is fine. This can perhaps be further explored through the sort of healthy living suggested earlier. At any rate, it is useful to emphasize food as being a source of energy so that we can move, grow and carry out all the other life processes on this list. The fact that our food also provides us with vitamins, minerals and things that we need to stay healthy can also be explored.
7. Produce waste. At this stage no great detail is necessary here. If the children have looked at changes in materials, this can be linked to the fact that our bodies change the air we breathe and the food that we eat in ways to make it useful to us but that this also makes new materials that would be harmful to us if we did not

get rid of them. Freudian key stage on children will take delight in focusing on going to the toilet but it is worth reminding them that we also get rid of waste when we breathe out.

Once the children's provisional list is established, they can explore the ways in which their bodies allow them to carry out these processes. At this stage this is likely to be a fairly simple exercise not going much beyond, 'we have joints in our skeletons to let us move', 'we have lungs to breathe' etc.

Conceptual step two – extending this understanding of life processes to: first other vertebrates; then invertebrates; and finally to plants

Having begun by looking at life processes in their most familiar context, that of humans, it is now possible to examine them in the context of other organisms, beginning with those most similar to humans and then progressing to those most different from humans. This will allow the children to gain a better understanding of what it is that makes an organism alive.

Eliciting ideas

A useful way to begin eliciting the children's ideas is to ask them to categorize various objects and organisms as to whether they are living or not and to explain the reasons for their choices. It is important to include examples of the various groups that the children will focus on such as all five types of vertebrates, some common invertebrates and various plants. It is also important to include examples of the sorts of objects that children often wrongly categorize such as cars, fire and clouds, which are commonly viewed as being alive; and things like eggs and seeds that children often don't realize are alive. This exercise will help the children focus on and refine their provisional list of the characteristics of living things.

Be warned!

The lists in the last section give examples of where children often have misconceptions but it is important to understand why this is as it reveals the shortcomings of the traditional checklist of the characteristics of living things.

Fire illustrates this very well. Fires move; they grow; they need oxygen (and, in fact respire in a parallel manner to ourselves); they require nutrition (fuel); they produce waste; they reproduce (one child once told me that, 'baby sparks fly off the mummy fire and start growing on their own'); they may even seem to sense by responding to the wind or, as another child told me, 'fires don't like water and keep away from it; they just go to the dry places'. It is hardly surprising then that the children can be confused.

On the other hand, an object like a sunflower seed seems to do nothing on the checklist. Children have often told me things like, 'the seed is dead now but if you put it in soil and water it, then it comes alive'.

Challenging the children's ideas

The first step is to begin reviewing the children's ideas on what is alive and what isn't. Their checklist of the characteristics of living things will provide a starting point but, as discussed above, extra help will be needed to challenge some of the children's likely misconceptions.

Firstly you will have to help the children see that living things carry out these functions in ways that distinguish them from non-living things.

For example, fire may seem to reproduce by producing sparks that grow into baby fires but this is not at all the same sort of reproductive process that living things go through. Later, in key stage 2, it will be possible to introduce the children to such concepts as cells and DNA but these are likely to be beyond most key stage 1 children and so alternative strategies are required. I have found it useful to refer back to having discussed that humans' bodies are organized in special ways to help them carry out their life processes, for example: we have eyes to see; legs to help us move about; or lungs to breathe. Even young children can appreciate that fire does not have a body that is organized in this sort of way and this helps them appreciate that it is not alive.

Even where confusing non-living objects seem to be organized like this, there are ways of helping the children see that there are differences from living things. For example, one sharp year two child recently told me that his 'Mummy's car was like a bat because it's blind but can tell when it's going to crash into things'. The child was talking about parking sensors, which do work in a similar fashion to bats' echo location (though note that bats are not blind!). Cars have well-organized structures that seem to parallel human bodies such as wheels to help them move and parking sensors to allow them to detect objects, so it is understandable that children should be confused. However, the child in the example easily accepted that his mother's car wasn't alive because, as he put it, after a few scaffolding questions 'it was built in a factory', rather than, 'coming out of its mummy's tummy like a little bat'.

Other difficulties can be best dealt with by a thorough consideration of life cycles. Life cycles are fascinating for young children but they are also very important in helping them appreciate the difference between living and non-living creatures. All living things are part of a life cycle where organisms produce more of the same sort of organism, which in turn produce more and so on. Although sparks from one fire can produce another fire, fires are not part of a life cycle; originally the fire must have been started by something else such as lightning, a person with a match, or the Sun's rays being focused through a broken bottle.

Life cycles also help challenge the difficulties caused by seeds, eggs etc. If the children see that seeds, for example, are part of a life cycle, they will more readily accept that they are alive and that the life processes in them are simply dormant and awaiting the correct conditions to begin again. This concept of dormancy is an important one and it helps if the children can make links to such concepts as animals hibernating, or deciduous trees losing their leaves and slowing their life processes in winter.

Once these sorts of areas have been addressed, it is then possible for the children to examine other living things and see how their life processes parallel those of human beings. Conceptually, this is not particularly difficult but I have found it useful to start with examples of organisms that are similar to humans and then to progressively move to those that are most different from ourselves.

Where at all possible, and with due consideration to health and safety, it is best if the children can meet and study examples of the living organisms, as in Figure 20.1.

The hierarchy of progression that I use is:

1 Humans.
2 Other familiar mammals such as pets or farm animals. (The last school that I worked in had a learning partnership with a local farm, which provided excellent opportunities for this.)
3 Birds. (Often falconry centres are a great help here.)
4 Reptiles. (I used to keep a garter snake as a class pet.)
5 Amphibians. (The old chestnut of bringing frogspawn into the classroom is now discouraged for conservation reasons but I used to have a local wildlife ranger bring in frogs and toads on a temporary basis for the children to learn about.)
6 Fish. (No school should be without a well-maintained tropical fish tank.)
7 Common invertebrates. (Earthworms, snails, and woodlice are three of the easiest to work with.)
8 Plants. (Every classroom should have a variety for the children to look after, enjoy and learn about.)

Figure 20.1 Children studying giant African snails to learn about their life processes

Conceptual step three – understanding how living organisms are adapted to different habitats

Once the children are familiar with the different life processes and how they are exemplified in different living organisms, it is relatively straightforward for them to extend their knowledge to learning about how different organisms are adapted to live in their various habitats and to their being able to compare and contrast different habitats.

Eliciting ideas

I have found that one of the best ways of establishing the children's thinking in this area is to play a game I call *Mystery Beasts*. I usually begin with a whole class discussion of the various problems that living things have to solve. The children will phrase these in many ways but they can be reduced to a short list, each of which can be related to the life processes the children have already learnt about.

1 Finding a suitable home;
2 Finding something to eat;
3 Avoiding being eaten by something else;
4 Finding a mate.

Once this list, or its equivalent, has been established, I then introduce the children to a selection of interesting and unfamiliar animals and ask them to make up a story that involves the animal in doing all of the things on the above list. The key point is that the children must study the animal and obtain evidence to support what happens in their story. For example, if they are looking at a green beetle, they might decide that it lives amongst leaves in a tree because it would be well camouflaged there. Depending on the age and the ability of the children, and on the particular pre-assessment needs: such an exercise might be undertaken as a whole class, in groups, or as individuals; varying degrees of scaffolding might be required; and the exercise could be undertaken purely orally, or be written in simple sentences. Pictures can be used but my favourite method involves giving the children actual specimens, usually of invertebrates. These can be living, or preserved specimens. I have quite a collection of such things now but for those whose cupboards are filled with more normal things I have often found that the local museum or ranger service can help out with some splendid specimens.

Be warned!

These conceptual areas have few equivalents of the deep-seated misconceptions found, for example, in topics such as forces. Obviously the children may show much factual confusion but this is easily dealt with. However, watch out for the inevitable tendency of young children to anthropomorphize the creatures. I don't think there is any harm in practices such as the children naming their mystery beast but gently guide the children away from statements such as one I remember a child making about the eating habitats of a giant tropical longhorn beetle she was looking at. When asked what the creature ate and why, she said, 'It eats pizza because I like pizza and we're best friends'.

Hint!

Although at first sight it may seem potentially confusing for the children, I have often found it useful for them to work on different sorts of animal stories around the same time. They might look at fictionalized animal stories; at factual writing about animals; at anthropomorphized animal cartoons; and at extracts from wildlife documentaries. They can then produce their own fictional stories as well as the scientific, evidence-based predictions of the sort of exercise described above.

Challenging the children's ideas

After practice in the sorts of skills involved in the elicitation exercise described, there is no substitute for getting the children out and exploring real habitats, seeing the creatures that live there and using their science skills to predict and then gather information about how the creatures fit into their habitats. It is possible to do this in a myriad of ways. Sometimes they can be independent such as a 'minibeast safari' around the school grounds. Often though, a great deal of help can be had from local environmental and ranger services. Just a few examples of wonderful work on habitats I have seen young children undertake have been: pond dipping on a local nature reserve; rock pool rambles with marine rangers; and woodland studies guided by forestry workers.

Conceptual step four – developing simple taxonomic skills

At this level, taxonomic work should be kept simple with an emphasis on the children's own observations and ideas but it is possible to lay sound foundations for classification that the children can build on in key stage 2. However, in the children's previous work on life processes, they will have been introduced to simple scientific groupings of living organisms and I have found that, with a little practice, even key stage 1 children can differentiate between animals and plants and, within the category of animals, successfully distinguish between vertebrates (mammals, birds, reptiles, amphibians, fish) and invertebrates.

Eliciting ideas

The best activities for eliciting the children's ideas are also those that provide the first steps in the children's learning about classification and simply involve asking the children to sort things. Here the emphasis is not on following any recognized scientific taxonomies but on establishing how the children sort and classify objects, for example looking to see if they can do this on the basis of observable characteristics, rather than predictions as to what an organism might do; and checking that they can use their own classifications consistently. Suitable objects might include: pictures of living creatures (these are especially useful for checking the children's ability to assign organisms to the scientific groups listed above); sets of plastic minibeasts; collections of sea shells; or sets of leaves the children might have gathered themselves.

Be warned!

Here the problems tend not to be deep misconceptions but rather the necessity to practise skills, such as consistency in applying criteria; or, factual confusions such as thinking a spider is not an animal, or a tree is a 'tree' not a plant.

Challenging the children's ideas

So far as developing the children's classification skills is concerned, the secret is simply plenty of practice.

When it comes to the children learning the first steps of correct scientific classification, once more emphasis should be placed on their own ideas but through scaffolding and challenging, guiding these to simple taxonomic criteria. Care must be taken however as some of the traditional biological definitions, for example the cell types of animals and plants, may be inaccessible to children at this stage. I have found the following key features to be useful. Above all, the most important thing is not to simply try to give the children simple definitions but to develop this understanding through lots of concrete examples and discussion.

When distinguishing between animals and plants, young children tend to focus on a limited range of animal characteristics: moving, having eyes and making sounds, are all ones I find children list commonly. These are fine in many respects but when the children encounter more confusing animals such as sea anemones, they may get confused. Accordingly, I have found it useful to especially think about what makes a plant a plant, not an animal. Here, even young children, aided by practical investigation, can appreciate that almost all plants (the exceptions being parasites that are rather obscure for children of this age) need sunlight, water and air, which they use to make their own food (or, more simply, as some children may only be able to appreciate, to grow).

I often find that young children love learning new words and exploring where they come from and so you may find that using the technical terms vertebrate and invertebrate is appropriate but, at any rate, the main distinguishing feature for the children is that some animals have skeletons inside and some don't.

Within vertebrates, allow the children to come up with their own lists but ensure these are rigorously challenged. As well as the children's ideas, I have found it useful to ensure the following key features are recognized.

Mammals	feed their babies on milk produced by the mother.
Birds	have feathers.
Reptiles	have scaly skins. They can live in dry places as they, and their eggs (when they lay them) are waterproof (in other words they don't easily desiccate).
Amphibians	need damp places to live and breed as they and their eggs are not waterproof and so easily desiccate.
Fish	by and large, live in and breathe water (with a few interesting temporary exceptions, such as mudskippers and lung fish!).

Be warned!

Whilst it is certainly not advisable to over-complicate matters for the children, it can be equally dangerous to over-simplify. For example, think of the stereotypical 'primary school plant'! As not only drawn by the children, but often as illustrated in text books, this has filamentous roots; a relatively tall and thin stem; broad leaves; and a daisy-like flower. My classrooms have had anything up to one hundred different plants in them: spider plants; aspidistras; dragon trees; yuccas; ferns; pony tail plants; rubber plants; lots of different cacti; lots of different succulents; and various plants growing in the fish tanks. Not one of these has ever looked like that stereotype! Make sure you don't cramp the children's view of the wonderful diversity of life by careless and inappropriate simplifications.

21

Developing a more advanced understanding of life processes and living things

Introduction

In many ways, the conceptual steps outlined here are not so much new concepts but a deepening of the concepts outlined in Chapter 20. Just as was the case with the learning discussed in Chapter 20, the learning here, whilst having challenges, does not have the deep misconceptions associated with it that some primary science topics do. Accordingly discussion here is kept to those areas where there are likely to be more difficulties.

Where this topic fits in

All the conceptual steps discussed here are suitable for key stage 2.

Although the learning is not as complex as in some topics, there is a considerable breadth to cover and it is likely that the topic will need addressing on several occasions in key stage 2. Often it is useful to have a topic which focuses particularly on humans; an in-depth habitat study will be necessary; and often I have found that some extra, mini-topics themed on particular types of living things such as birds or trees, not only consolidate learning but form an excellent basis for cross-curricular work.

Links to other topics

Just as was discussed regarding key stage 1 in Chapter 20, science skills should be at the heart of the children's learning but, in terms of subject knowledge, this topic tends to stand alone.

A context for the topic

Once again, the children tend to be very interested in the topic of living things but it is still a good idea to plan for meaningful contexts that focus the learning and give the children a motivating reason for their work. There are many options but some examples regarding the topics mentioned will illustrate the sort of thing to aim for.

In my experience, healthy living topics can be rather dull for the children but, with some imagination, they can be made much more interesting. For instance one upper key stage 2 class I worked with became life style coaches and ran a special evening where their parents and other folks from home could come to the school and undertake various activities from having their pulse measured to healthy cookery demonstrations.

The value of getting the children to participate in actual scientific studies of living things was mentioned even for key stage 1 children; at key stage 2, this applies even more strongly and is an excellent means to develop the children's science skills, as well as their knowledge and understanding. With guidance from local rangers, or environmental organizations, the children can do some serious and useful work. In one of the schools I worked in, the older children did regular seabird counts, as part of the recording system for the area. Other projects have included: surveying dog whelks, as a means of monitoring marine pollution; building and monitoring nest boxes on local farms; woodland regeneration projects; and surveying butterflies.

Conceptual step one – extending knowledge of life processes to more complex areas

This conceptual step is really just an extension of the equivalent work in key stage 1 with a more technical emphasis being placed on the three linked areas of respiration, excretion and nutrition. It is particularly important to pay careful attention to the

Table 21.1 Example learning objectives for developing a more advanced understanding of life processes and living organisms (all of these conceptual steps are suitable for coverage at key stage 2)

Conceptual step	*Example learning objectives*
	Children will learn:
Extending knowledge of life processes to more complex areas.	that the life processes of living organisms are movement, reproduction, sensing, growth, respiration, excretion and nutrition; how these life processes are exemplified in a range of vertebrates, invertebrates, plants, fungi and bacteria;
Appreciating the complex web of adaptations and relationships in different habitats.	the relationships that link living organisms in different habitats; how the life processes of the types of organisms identified in conceptual step one shape how they relate to their habitat and the other organisms within it;
Using keys to identify organisms and develop simple taxonomies.	to create keys to differentiate between living organisms; to use keys to identify commonly encountered organisms within habitats that the children are studying; to assign living organisms to major taxonomic groups.

addressing of these areas regarding plants, as that is one area where more entrenched misconceptions are likely for this topic.

Eliciting ideas

In many respects, the children will begin by more or less just revising the concepts they will have encountered earlier and I find an effective way of pre-assessing their ideas is to present them with examples of some familiar and less familiar living organisms and ask them to predict what their life processes are and how they carry them out.

It may be useful to check for any lingering misconceptions by including some non-living examples and some of dormant stages of life cycles, such as seeds.

It is also vital at this stage to ensure that examples of plants are included, since, as was discussed in Chapter 20, children are often confused by plant nutrition.

In key stage 2, I normally introduce the children to two further taxonomic groups of living organisms: fungi and bacteria. It might therefore be appropriate to include examples of them in the pre-assessment. This will firstly allow you to check if the children regard them as living organisms; and secondly allow you to see the children's ideas on what life processes they may, or may not have.

Be warned!

The children may show evidence of persistent misconceptions of the sort that were discussed in Chapter 20 but, with one exception, any confusion encountered at this stage is most likely to be factual, rather than of a serious conceptual nature.

The exception is regarding plant nutrition. In the previous chapter it was pointed out that children have difficulty in accepting the fact that almost all plants actually make their own food using the energy in sunlight to build sugars out of water and carbon dioxide. The technicalities of photosynthesis will almost certainly be beyond such young children and so, at key stage 1, I simply emphasize that almost all plants need sunlight, water and air to be able to live. I would usually explain that this is because such plants can do something amazing that no animal can do; in other words, make their own food. However, I accept that many of the children may not successfully understand this concept and that in key stage 2 they are still likely to hold the classic misconception that plants get their food from the soil. Often the children will have a sophisticated model for this and explain that the food gets dissolved in water and taken up through the plants' roots; a misconception reinforced by the fact that many fertilizers are soluble and are often unhelpfully termed 'plant food'.

This misconception also highlights a factual point that it is important to realize is likely to confuse children; the nature of respiration and how it is the process that releases energy from an organism's food.

Challenging the children's ideas

It is useful to begin with a discussion of why we, as humans, eat.

After their work on humans as living organisms, I find that children quickly appreciate that there are two aspects to their food. The first is that it provides the energy needed for them to move, grow and carry out all their other life processes. The scientific concept of energy is one which I find largely beyond all but the most capable primary school children. In Scotland and Northern Ireland, where energy was considered a primary school topic, I found that the children could learn facts about energy but had little serious conceptual appreciation of what it was. Hence there has been no mention of energy even in the chapters on electricity, light and forces, where, perhaps, it might have been expected; instead, less abstract, mechanical models are employed in helping the children understand the concepts. However, the one area where energy is conceptually accessible to the children is regarding life processes. I think this is largely because the everyday use of the term 'energy' in this sense closely mirrors its scientific equivalent; the children appreciate that without their food they would, as one child put it to me, 'be sort of all tired and not be able to run about or anything'. The second aspect is that, as well as energy, food provides us with things we need to stay healthy such as vitamins and minerals. Even at primary level, the children might learn what, for example, some of the main minerals and vitamins are and why we need them, such as: calcium for strong bones and teeth; or vitamin C for healthy gums and to help fight infections. However, even if these substances are vital for us they do not provide us with any energy; a pure diet of vitamin and mineral supplements would soon leave us, 'all tired and not able to run about or anything'.

Once the children have appreciated this distinction, they are much more likely to accept that the dissolved nutrients that plants typically absorb via their roots are the plants' equivalent of vitamins and minerals and that, however important they might be, they do not actually provide the plant with any energy. Instead, plants get their energy from food they make themselves using the process of photosynthesis.

Unfortunately, there is no direct manner in which photosynthesis can be demonstrated to the children. However there are a number of activities that can indirectly help the children's understanding of the process.

1 Firstly, reference can be made to children's understanding of chemical reactions, as discussed in Chapter 9. The children will be familiar with how chemical reactions involve substances combining to make new substances. Photosynthesis is exactly that: water and carbon dioxide are combined and turned into sugar with oxygen being produced as a waste product.
2 The importance of light in providing energy for the reaction can be shown by simply keeping a seedling in the dark. The plant may use stored food to survive for a little and may even show a burst of growth as it desperately seeks for light but it will soon yellow and die if not restored to the light.
3 Children can be introduced to the role of chlorophyll (the green compound in plants) in photosynthesis. This can be indirectly demonstrated by covering parts of a plant's leaves with an opaque substance; the green chlorophyll will be moved away from the shaded area to where it can be more useful, leaving the shaded area of the leaf yellow in colour.
4 The fact that a gas (oxygen) is produced as a waste product can be shown by

keeping some pond weed under the cut off, top section of a sealed pop bottle, that is stuck to the bottom of a tank with adhesive putty. The tank, and the pop bottle, will be initially filled with water but, as the plant photosynthesizes and gives off oxygen, a bubble of gas will appear at the top of the bottle section.

Such work then lays the foundation for a more advanced understanding of what respiration actually is.

In key stage 1, it is appropriate to think of respiration simply as the fact that all the living things that children are familiar with require air, or, more specifically, oxygen. Often children at that stage talk of living things as all 'breathing', even if they realize that a human, a fish and a plant all do this in different ways. However, respiration is not just breathing; rather it is the chemical reaction that allows living things to release the energy in their food; a chemical reaction that requires oxygen, hence the need for breathing, or its equivalent. This concept can be introduced at key stage 2, even if the complexities of the reaction need not be discussed. The children can also draw on their knowledge of the chemical reaction of burning, which is an exactly parallel reaction, and also requires oxygen.

Conceptual step two – appreciating the complex web of adaptations and relationships in different habitats

Once again, the children will largely be extending the concepts they have learned in key stage 1, rather than encountering completely new ones. In key stage 1, the focus was on simple adaptations of living organisms, largely to do with finding food and protecting themselves. Now more complex relationships can be explored, such as parasitism, and the exploration of the relationships can be extended to include such concepts as food chains and webs.

Eliciting ideas

By far and away the best way for children to learn about living organisms, about their adaptations and about how they interrelate in various habitats, is to actually go out and study them. Accordingly, I always aim to use field work as both a means to motivate the children at the beginning of a topic and as a way to raise questions that can be used as a means of pre-assessing the children's ideas. An initial visit to a habitat to be studied can be arranged and the children encouraged to make observations of what they think is going on in the habitat and to raise questions. These observations and the children's predicted answers to the questions will give you considerable insight into their thinking and understanding. Some examples will help illustrate the process.

One example is when beginning a woodland study, where I wished the children to ultimately learn about food chains and webs, I asked the children why they thought we saw more chaffinches than sparrowhawks.

Another example involves visiting a local wetland reserve. Any study of the natural world will quickly lead to observations that allow you to elicit the children's ideas on how organisms are adapted to their habitat. On our initial visit to the wetlands,

I encouraged the children to think about why there are such marked differences in the plumages of male and female ducks; but why male and female geese and swans have identical plumages.

I have found that the children are fascinated by these sorts of questions and, as well as allowing you insight into their ideas, the questions then form a great springboard for practical, skills-based work, and research.

Hint!

I always plan to have the children undertake long-term studies at some point in their time in primary school. For example: in one school we did a beach study where we visited the beach and examined the habitats in the various seasons; whilst in another, the topic of plants was actually studied as four mini-topics, looking at plant life in spring, summer, autumn and winter. In studies such as these, different questions will be raised at different seasons. Sometimes you may wish to scaffold particular pre-assessment questions such as: why do most woodland flowers blossom in the spring? Or, why are most berries red, or a shiny purple/black in colour? Sometimes the children themselves will raise the questions, such as the children who noticed that when we visited the rock pools in winter, the winkles seemed to be clustered together, rather than scattered around singly, as they had been in early autumn, and asked why this was.

These examples also show how the pre-assessment questions make a great foundation for subsequent, skills-based, investigative work.

For example, regarding the spring flowers, some children correctly predicted that the woodland plants could only grow and flower properly when they were not shaded by the full canopy of leaves of the trees above them. These children then set up a long-term investigation to see if the flowering time of plants varied depending on the amount of light they got. Other children thought that insects would not visit the plants if they were in shade, another excellent prediction. They then monitored test plants standing in the Sun and compared the insects visiting these with other, identical, shaded plants.

The winkles show another excellent example. Here the children suggested that the clustering was to stop the winkles losing heat so quickly when they were exposed to the air at low tide. Those children had done the sort of work discussed in Chapter 10! The children had then to design a sampling strategy that allowed them to check if there was a correlation between temperature and winkle cluster size; a correlation that they did, in fact, find.

Be warned!

The good news is that, especially if the children have undertaken learning in the fashion that was discussed in the previous chapter, you are unlikely to encounter deep misconceptions, though the children will, of course, have some factual confusion.

The one area where I have found a consistent, though relatively easily challenged misconception, is regarding food chains. Many children do not initially realize that there must be a much greater amount of plant material than there are herbivores that eat it and that, likewise, the herbivores must outnumber the predators.

Challenging the children's ideas

I think that the major reason that the children get confused over the hierarchy of food chains is that there is a positive correlation between an organism's position towards the top of the food chain and how it sticks in a child's imagination! Thus a top predator, like an eagle, seems much more significant than a tuft of grass, despite the fact that the eagle (and virtually every other living thing on the planet) is dependent on the ability of plants to harness the energy in sunlight to make their own food, which can then be eaten by herbivores and so passed up food chains. This means that food chains actually form food pyramids: there will always be fewer predators than prey; and always less herbivores than plant material. The technical reason for this is that energy is lost at each stage of the food chain but I am content if the children come to an understanding such as, as one girl recently told me, 'if there were too many caterpillars, they would eat all the plants and then they would just be hungry and die'.

Conceptual step three – using keys to identify organisms and developing simple taxonomies

The children can now build on their key stage 1 work of sorting organisms according to their visible characteristics and begin to make their own simple keys and to use keys to identify organisms they encounter in their studies. In addition, they can extend their knowledge of scientific taxonomies to include fungi and bacteria, and to add more detail to their classification of invertebrates and plants.

Eliciting ideas

The best way to pre-assess the children's understanding of keys is to ask them to construct simple keys of their own. I often use pre-prepared sets of objects that relate to the topic being covered. For example, if we were learning about minibeasts, I often give the children a selection of plastic minibeasts to make a key for; if we were doing a winter plant topic, I might give them sets of winter tree twigs to key. The advantage of working in this way is that it is possible to select the objects to key to take account of the children's likely ability. Often it is useful to begin such exercises with some whole class, or group discussions of the task to orientate the children (see Figure 21.1).

When it comes to time to check the children's ideas on actual scientific classifications, the best activity is to give them pictures of the range of organisms to be classified and to ask them to sort them into groups and justify their choices. Care must be taken however that the pictures chosen clearly illustrate the relevant taxonomic characteristics.

Often a good way to start is to secretly pick a minibeast and then for the children to work out which beast you have picked by asking questions that can be answered yes or no, and that will allow the group of objects to be progressively whittled down until only the picked minibeast is left. There are a wide variety of plastic minibeasts available and, although the colours may be garish, and they may not be to scale, with a little care, it is possible to make collections that are actually anatomically correct, which will help the children with their taxonomic understanding.

Figure 21.1 A preparatory discussion with a group of children before they design a key to differentiate some plastic minibeasts

Be warned!

Deep misconceptions are unlikely here but there can still be some problems.

When designing keys you must always encourage the children to base their keys on clearly observable characteristics. Often I find that the children try to use prior knowledge as a basis for their key questions. For example, the sets of plastic minibeasts I use have bees and wasps in them and children regularly come up with key questions such as, 'Does the minibeast have a sting?' This defeats the purpose of the key: the sting is not a visible characteristic of the models and, if the key user can identify a bee, or wasp and know that it has a sting, then she doesn't need the key!

When it comes to classification, you should take care in deciding how detailed a knowledge of scientific taxonomy you wish the children to learn. It is actually not that difficult to drill the children in learning categories but this is only meaningful if the children understand the distinctions and are using them as part of skills-based science work.

Challenging the children's ideas

With a little practice, keys are not actually too difficult a skill for primary children to master. I was initially surprised when I found that for most children, designing keys is easier than using them. It makes sense however, in that, when the children design their own key, they are in control of the questions and fully understand how they relate to the evidence. On the other hand, when the children are using a key, they often have difficulty interpreting the questions and may have difficulty observing the necessary characteristics on actual specimens. This means that great care must be used in selecting published keys for the children to use, otherwise poor identification and considerable frustration can result. As a general rule, the easier the key is to use, the less likely it is that it will allow the children to identify all the organisms that they might encounter, so there is a balance to be struck. I often designed my own keys for the children, based on the particular habitats I was taking them to but, if you don't have the experience to do this, a chat with local wildlife rangers or environmental organizations will often bring you good advice and ideas.

When it comes to expanding the children's understanding of scientific taxonomies, you must judge the level of complexity that your children are likely to cope with. The key point is that there is little benefit in having the children learn classification as a collection of de-contextualized facts; it should always be linked to the children's science skills such as observation and recording, as they study habitats and learn about the organisms that live in them. For instance, just learning the correct taxonomic terms, mollusc, annelid etc., is a fruitless exercise. These should always be linked to the underlying concept of classification by observable characteristics. Often this can be reinforced by some useful literacy work on word derivation, which can be illustrated by the two examples just given. A major defining characteristic of molluscs is a soft body and the term mollusc comes from 'mollis', the Latin word for soft. The major defining characteristic of annelids is a body divided into ringed segments – annelid comes from 'annulus', the Latin word for ring. Likewise, teaching the old chestnut that insects have a 'head, thorax and abdomen' is pretty meaningless if left at that. If, however, you link this to observational skills and comparisons with other types of minibeasts, it becomes a useful learning exercise.

Hint!

When designing keys there are many great, primary level ICT packages for branching databases that will not just help the children's science skills but will also enhance their ICT skills.

Further reading

Your own subject knowledge

These titles are useful for deepening your own science subject knowledge. They are useful but care is needed as they often enter into both types of explanation and levels of explanation that may be appropriate for you, but will be beyond primary school children.

Deveraux, J. (2000) *Developing Primary Science Subject Knowledge*. London: Paul Chapman.

Farrow, S. (1999) *The Really Useful Science Book: a framework of knowledge for primary teachers*. London: Routledge Falmer.

Johnsey, R., Peacock, G., Sharp, J. and Wright, D. (2009) *Primary Science Knowledge and Understanding*, 4th edn. Exeter: Learning Matters.

Wenham, M. (2005) *Understanding Primary Science – Ideas, Concepts and Explanations*. London: Paul Chapman.

General primary science

These titles are useful in helping you develop your practice in general areas of primary science education.

Asoko, H. and de Boo, M. (2002) *Analogies and Illustrations: representing ideas in primary science*. Hatfield: ASE.

Harlen, W. and Qualter, A. (2004) *The Teaching of Science in Primary Schools*. London: David Fulton.

Hollins, M. and Whitby, V. (2001) *Progression in Primary Science: a guide to the nature and practice of science in key stages 1 and 2*. London: David Fulton.

Ollerenshaw, C. and Ritchie, R. (1997) *Primary Science – Making it Work*. London: David Fulton.

Roden, J. (2005) *Reflective Reader: primary science*. Exeter: Learning Matters.

Sharp, J., Peacock, G., Johnsey, R., Simon, S. and Smith, R. (2009) *Primary Science: teaching theory and practice*. Exeter: Learning Matters.

Sherrington, R. (ed.) (1998) *ASE Guide to Primary Science Education*. Hatfield: ASE.

Science skills and investigations

Although the major focus of this book is the conceptual progress of the children's knowledge and understanding, the methodologies outlined in the book are based on eliciting the learners' ideas and challenging and progressing them in a skills-based manner. These sources will help you develop your understanding of science skills further.

Ball, S. (2001) *Speak to Me Graph.* Hatfield: ASE.
Feasey, R. and Goldsworthy, A. (1997) *Making Sense of Primary Science Investigations.* Hatfield: ASE.
Goldsworthy, A. (2003) *AKSIS Investigations: making an impact.* Hatfield: ASE.
Goldsworthy, A., Watson, R. and Wood-Robinson, V. (1999) *Getting to Grips with Graphs.* Hatfield: ASE.
Goldsworthy, A., Watson, R. and Wood-Robinson, V. (2000) *Developing Understanding in Scientific Enquiry.* Hatfield: ASE.
Keogh, B. and Naylor, S. (1997) *Starting Points for Science.* Sandbach: Millgate House.
Wenham, M. (2001) *200 Science Investigations for Young Students. Practical activities for science 5–11.* London: Paul Chapman.

Health and safety

Safe practice is vital and these titles are an excellent start towards ensuring it.

Abbott, C. (ed.) (2001) *Be Safe! Some aspects of safety in school science and technology for key stages 1 and 2.* Hatfield: ASE.
Wray, J. (ed.) (1994) *Safety in Science for Primary Schools: an inset pack for use by head teachers, advisory teachers and teacher trainers.* Hatfield: ASE.

Children's ideas in science

These books provide further information on children's ideas within the specific science topics addressed in the book.

Driver, R., Guesne, E. and Tiberghien, A. (eds) (1989) *Children's Ideas in Science.* Buckingham: Open University Press.
Hollins, M. and Whitby, V. (2001) *Progression in Primary Science: a guide to the nature and practice of science in key stages 1 and 2.* London: David Fulton.
Naylor, S. and Keogh, B. (2000) *Concept Cartoons in Science Education.* Sandbach: Millgate House.
Osborne, R. and Freyberg, P. (1985) *Learning in Science: the implications of children's science.* Auckland: Heinemann.

Index

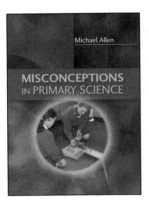

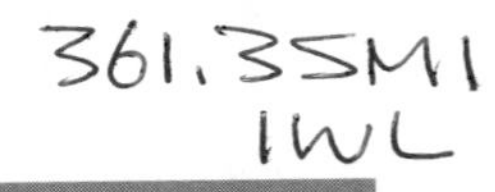

Families, Carers and Professionals